Hybrid Humans

Dispatches from the
Frontiers of Man and Machine

Harry Parker

PROFILE BOOKS

This paperback edition first published in 2023

First published in Great Britain in 2022 by
PROFILE BOOKS LTD
29 Cloth Fair
London
ECIA 7JQ
www.profilebooks.co.uk

Published in association with Wellcome Collection

**wellcome
collection**

183 Euston Road
London NWI 2BE
www.wellcomecollection.org

10 9 8 7 6 5 4 3 2 1

Typeset in Garamond by MacGuru Ltd
Printed and bound in Great Britain by
CPI Group (UK) Ltd, Croydon, CRO 4YY

A CIP record for this book can be obtained from the British Library

ISBN: 978 1 78816 311 8
eISBN: 978 1 78283 583 7
Audio ISBN: 978 1 78283 991 0

For Caro

Contents

Dreams of a Broken Body

I open my eyes to the ward. The soundtrack is back: the low hum of machines, the shuffle of a nurse, a monitor chiming on the other side of the ICU. I inch my head around. Curtains are drawn around the bay and the lights are dimmed. The middle of the night. I lie and stare at the wall and become aware of my body. A roll call of body parts checking in. It's too painful and strange to be me – I am dislocated from this broken flesh. My nerves have been shocked by explosives, and everything below the neck is fizzing. There is a whirring also: the analgesia dulling the pain, retuning its frequency so it is just white noise. I feel for my legs along the map of synapses I've known a lifetime, but now my legs are distant, shimmering in a kind of hot furnace beyond my leaden arms and aching back and through the sharpness of lesions and bruising and the fizzing white noise blossoms into pain.

I recoil from it and press the PCA button; it drones and

pushes morphine into my central line and after a while I am unconscious.

During those first weeks the shock and drugs and countless surgeries distorted everything. Dreams crossed into waking, so nothing felt real. My imagination seemed to be protecting me from what had happened, walking me around the ward so I could look back at the bed and my body in it, or transporting me to places I knew from childhood: to dreams of home, of schools, of the town centre where I was first given the freedom to go shopping on my own – walks from my youth that felt more real than the medicalised world I'd woken to. It was as if my imagination had kicked into overdrive, emerging to take the reins and guide me through the trauma and strangeness – to help me accept a body I knew was mine, but which was full of pain and attached to the wall of a hospital by pipes and wires. And broken. The left leg gone below the knee, the right halfway down the thigh.

I've fumbled around for the memory: that moment I gasped wide-eyed with the realisation that I had lost my legs. I can't find it. No appalling shock of a doctor or family member breaking the news to me. Instead there are many wakings – from deep comfortable sleep, from anaesthetic oblivion, from dreams horrible and brilliantly surreal – each one eroding a little more of my old self, making way for the new one that was forming.

Each year, as 18 July approaches, I think again of how my life has changed. I've heard American military veterans call it a 're-birthday' and celebrate their second lives. I did mark the day in the first year. I had a load of friends round to my flat and had a barbecue. Just a party, no speech or cake, and most people didn't let on that they knew the significance of the date. As the years have passed, I've done less: a night in the pub or a raised glass over dinner. And last year I received a text at lunchtime from a friend telling me to *have a good one* and it took me a few seconds to realise what he was talking about.

The moments of stepping on an improvised explosive device (IED) are seared into my memory – they seem as perfectly formed as the day it happened: unforgettable, yet probably as unreliable as any of my memories, altered and embellished with each re-remembering. I don't think about it much any more. It's been ten years, and too many experiences are stacked between that day and now. My dreams are different. In sleep, I don't see myself with or without legs, I simply see me. And the daydreams of that broken body lying in hospital have changed too: it is neither what I hoped for, nor what I feared – it is normal. A loss grieved for and accepted. I am not a victim, unable to walk, nor am I entirely freed from my disability. And while some horizons have contracted, others have expanded. Now, if I was offered the chance to rewind, to never have stepped on a bomb, not only would I refuse, I'd actually be terrified of losing this new part of my life. It would be to change my identity, to erase all those experiences, both good and bad, that make me who I am.

It is 18 July again. I am reminded this year by the date on the appointment letter from the limb-fitting centre (my microprocessor knee needs a service). I'm late. It's been a hard night. Our two children have taken turns being awake – another cold – and now my daughter is refusing to get dressed. I pull on my legs while discussing with my partner who will do the nursery pick-up. It is automatic: roll the liners over my stumps and click into the sockets, no novelty, no flinching – a muscle memory strengthened over ten years. Then I put in my contact lenses. The first goes in routinely but my eye flinches around the second lens, flattening it against my finger. I try prodding it in again. It falls on the floor.

In my early thirties I started wearing glasses. I hated the way they felt and the barrier they threw in front of the world, so I tried contacts. 'Only a few hours each day to begin with,' the optician had said, 'let your eyes get used to them. Let your tolerance build up. And take them out in the evenings, so your eyes can rest.'

'Like learning to use prosthetic legs,' I replied, but he didn't understand.

My half-dressed daughter is banging the shower door open and closed. It's a spaceship and she's going to the Moon. She wants me to come too, but I'm trying to find the lens. There it is. I peel it off the floor, clean it in my mouth, then push it in. My eye waters with pain as I persuade her out of the spaceship and downstairs. I need to find my leg's remote control for the appointment. Through the blur of tears I see it in the key pot, then apologise for the breakfast-time chaos I've abandoned my partner to and leave.

During the drive to the Central London hospital I keep rolling my finger over my eyeball, trying to dislodge whatever is behind my contact lens. I'm pretty sure I've scratched my cornea now. I crane upwards to the rear-view mirror. My eye is bloodshot and closing around the irritation. It's distracting and hard to drive, so I pinch out the lens and flick it away. I will spend the rest of the day with half the world drawn in a misty haze. For once my legs aren't the most annoying med-tech I use.

I'm stuck in traffic and look through the myopic blur at the people on the pavements. The school-bound children are running and jumping onto the brick edging of a flowerbed, bouncing into each other and laughing. They fizz about like loose atoms among the older pedestrians. I notice almost all the adults making their way down the street have a slight limp, an asymmetry to their gait or glasses, or one shoulder lower than the other. Further on there is a man on a mobility scooter. Bodies losing the suppleness of youth, and ageing.

I look for the technologies used to delay, rebuild or replace these losses of youth. How many of them have popped a pill this morning for an illness or pain, or to enhance their diet, mood or intellect? A woman is shuffling past my car now, rotating her waist around a walking stick. Hip transplant probably, or on the waiting list for one. There's a woman in a trouser suit hurrying through the crowd. I imagine a pacemaker keeping her heart in time. The children are gone, skidding around the corner, school bags wheeling.

A bus stops beside me. I look at a teenager sitting on the bottom deck. His neck is bent to a phone, his shoulders hunch to it, white pods in his ears. Suspended in his own reality.

Being an amputee in the twenty-first century doesn't make me an outlier; we are all hybrid. And we all suffer losses. For some it is the loss of youth; for others it will be more profound. The possibilities to replace that loss – to merge human and machine – are greater than ever before. Artificial hips and knees are prolonging mobility, stents and shunts are increasing lifespans, retinal prostheses and cochlear implants are enhancing impaired senses. And as technology improves, so the likelihood of using a prosthetic, orthotic, implantable or wearable during our lifetime increases.

I drive on. Whatever was blocking the road has cleared and I feel the vibration of the accelerator pedal through my prosthetic.

I'm at the frontier. I'm with the pioneers.

Becoming Hybrid

If I had stepped on the IED that injured me while walking across the car park of a Central London hospital there's a small chance I would have survived, but my chances were actually far better where it happened, 4,000 miles away in Afghanistan, ten years previously. Despite being in the goat-shit-laced dirt of a small patch of irrigated desert under the unforgiving heat of a twenty-first-century war, there was no better place on Earth to sustain my injuries. Within eighteen minutes I was delivered to Camp Bastion field hospital, the best trauma hospital in the world, with just enough life left to be saved.

It was a dramatic and unusual way to become disabled – an origin story that sets me apart and very nearly resulted in my death.* I've attempted to make sense of those moments

---

*In those moments I became a member of a community that was

7

when my survival hung in the balance, to replace the half-remembered personal myth with something more truthful.

I remember the shouts, the noise of the helicopter and the pain pressing on my chest. Being a small body crushed in Death's massive fist. I felt people working hard to save me, but don't remember seeing them. Eyes screwed shut. Teeth gritted. It was an interior universe of agony and terror – a very painful race against oblivion. The overwhelming feeling: that my experience was shrinking to a pinprick, and I had to fight that shrinking with every fibre of my being; if I didn't, I would die. But what really saved me were a complex set of interdependent twenty-first-century technologies and the people who knew how to use them. Before I became surrounded by the assistive technologies of disability there were the technologies of survival.

I met the CMT (combat medicine technician) only a few months after my injury. We were having tea in a red-brick mess in an army barracks. She was presented to me by people who thought it might be a nice moment. They watched on, intrigued: *This person saved his life – what will he say?* What do you say to the person who brought you back from the brink? I could feel everyone watching, sipping their tea.

'How are you, Corporal B?' is all I could manage.

---

far larger than I had ever realised. In the UK one in five is disabled – that's eleven million of us. Only 17 per cent of this number are born with a disability; the rest acquire one through the wear and tear of ageing, a critical illness or sudden traumatic event.

It was awkward, and I was embarrassed and said something about how we had all done our jobs.

On the morning I was injured, Corporal B was patrolling a few yards behind me, carrying her backpack filled with saline, chest drains, chemical blood-clotting powder, tourniquets and dressings. Some of the most useful lessons we learn from war are medical. First among these is to treat trauma as early as possible, to push the right people and equipment as far forward as we dare, all the way up to the front line. She was at my side seconds after the explosion: patching up, tightening tourniquets, opening my airway.

But I was still losing blood from multiple wounds, I'd stopped breathing, shock had set in and I needed to be out of that exposed field. The 'Golden Hour' is a pillar of emergency medicine. People who suffer major trauma are more likely to live if they receive definitive care within sixty minutes of injury.* During the First World War injured soldiers could wait days in flooded shell holes, lucky if they lived long enough to see stretcher-bearers arrive to take them back to the field station. In the Second World War we reduced the wait to ten hours; during the Korean conflict, to five hours; in Vietnam, to one hour. It is the helicopter that annihilates time and space. The helicopter that flew me across the desert

---

*The American doctor R. Adams Cowley was central to the establishment of the 'Golden Hour'. His work during the 1960s formed the basis of trauma medicine worldwide. He also invented a prototype pacemaker that was used by Dwight D. Eisenhower and has a clamp for open-heart surgery named after him.

collapsed ten hours of tortuous driving through minefields into a few minutes' flight.

Years later I met a MERT (Medical Emergency Response Team) helicopter pilot in a pub – a friend of a friend. We chatted over beers, slowly separating from the group at the bar by the intensity of what we shared. The more he talked, the more he looked hollowed out by it. He eyed me up, trying to work out if I'd been one of his – as if he might find some redemption in my survival, a shred of meaning that he could buttress against *Was it worth it?*

The MERT carries a trauma team (two paramedics, an anaesthetist and an emergency medicine consultant) and all the drugs, equipment, ventilators and monitors they need to the point of injury. The casualty is stretchered through the dust, up the ramp, and placed on the floor of the helicopter. The team surrounds the patient and tries to save them. They get bloods and fluid in, aggressively pack wounds, clear the airway and intubate, dull the pain and stabilise. All this while the airborne emergency bay pitches this way and that, and the racket drowns out all but the loudest shout. In more than nine years of conflict MERT picked up thousands of casualties and started their treatment while they were still flying across the desert.

More recently I met a doctor for a coffee to discuss a project. I thought we'd never met before – he was a surgeon at an inner-city trauma centre, patching up kids who had been stabbed. We settled with our mugs, and he pushed a plastic pocket with a grey photocopy inside towards me. I pulled it closer. The contrast was poor: a greyscale reproduction of pencil on coloured paper, in a doctor's script.

'What's this?'

'It's a note I made on the surgery you had when you first came into Bastion. I thought you'd like to see it.' He pointed a finger at where 'RIGHT TURN' was written and underlined. 'You were the first *right turn* – the first to go straight to the operating theatre from the helicopter. It saved time. We realised there was no point in putting you guys through resus – you all ended up in surgery anyway. After you, it became standard procedure; we even do it in the NHS now.'

'I didn't know you were there,' I said.

'Sorry it's such a bad photocopy,' he said.

I read through some of it and he helped when I couldn't make it out or didn't understand the doctor's code. *Left below-knee amputation. Some soft-tissue loss mid-calf; bone loss at distal tibia (tourniquet in situ). Right leg – massive posterior soft-tissue loss (tourniquet in situ mid-thigh). Frag. injuries left and right arms; scrotal frag. with loss of left testicle.*

'What does that say?'

He looked and read the sentence. '*Laparotomy conducted due to precipitous drop in blood pressure to fifty-five systolic during final stages of the debridement* – I had to open you up, right at the end when we thought you were through the worst, your blood pressure dropped – and fifty-five systolic is pretty low.'

Running vertically down my stomach is a wide scar. Six inches long, flanked by rows of little white bump staple-marks, it skirts around my belly button and towards my groin. I'd never known why it was there – just another scar to go with the rest.

'It was a bit of a rush,' he said.

This man had confidently cut me open, flopped some of those organs out, had a good look inside, packed it all away and then stitched and stapled me back up.

These people weren't overwhelmed into indecision. They were playing at the boundary of human capacity for loss, and testing what might be possible. By the end of 2009, if you made it to the field hospital in Bastion with even a hint of life, the chances of dying of your wounds had dropped to 1.8 per cent. If you had a massive blood transfusion, as I did, the chance of death was 4.8 per cent. In a civilian trauma centre that figure would be nearer 30 per cent.

One moment I was gritting my teeth against the pain and thinking I would die, the next I woke to a hospital in Birmingham, alongside other unexpected survivors brought back from conflict.* When I was well enough, I was pushed from intensive care to join them on the wards. They were in various states of repair – bound with dressings, eyes patched over, limbs in external fixation cages, hands sewn into abdomens to keep the flesh alive – among a forest of drip stands and monitors, shiny helium get-well balloons and empty paper McDonald's bags.

But the unspoken question of many who visited the ward, glancing around at the more seriously injured was: *Wouldn't*

---

*If I had been injured even one year earlier, I would not have survived. All the various practices, techniques and technologies weren't yet in place.

it have been better if you hadn't survived? The point was made differently, sometimes about my own injuries.

'I'm not sure I could go on, if it happened to me,' one of my visitors said. He added a compliment: 'You're doing so well, you're a real inspiration.' We were a handful of soldiers who had survived when we shouldn't have – that story was written on our bodies – and he couldn't understand it.

Those who wrung their hands in pity made me angry. Implying my body was too broken to be worthwhile seemed to be an inability to imagine how someone might be capable of adapting. Their sadness also hinted at something more deeply rooted: that we struggle to separate the body from our belief in what makes us human – a body has a shape that is normal, and there are losses that seem so great they cannot be endured.

They couldn't see that the story was only just beginning. The transformation we had to make lay ahead of us – medicine and technology would fix us. But in those early days we too found it hard to imagine, and the jokes and mocking we dealt each other across the hospital ward concealed both sadness and uncertainty. Physical injury could be endured; it was harder to come to terms with what this loss had done to my sense of self. And I lay in a room of unexpected survivors, wondering what my future would be like. What kind of human would I become?

The surgeons had closed up my body with stitches and skin grafts and sutures – made it into a new, smaller shape, one that could still support life. I had waited in hospital as my body healed, the infection withdrew and the nurses and family kept me alive with drugs and care, and love. I had survived, I had acquired a disability, now I would find out how disabled I would be.

When the doctors were happy, I was transferred to the military rehabilitation centre. Had you met me in the corridor during those early weeks of rehab, there was everything that you would notice – the missing legs, the dressings stained with yellow discharge, the dark shrapnel pockmarks up my forearms, the skinny, hunched body in the wheelchair, the tangy whiff of wounds and medical fluids – and there was everything you would not see: the pain of injury; the buzzing in my neck that no one could explain; the fog of sixteen colourful pills twice a day; the embarrassment that *this is me now*. And, most deeply hidden, the jarring distress I suffered when what I expected to feel, the experience of being me, no longer matched my reality.

I felt shrunken. Healing had consumed my fat and muscle and made my arms stick-thin and my ribcage prominent. Everything was out of reach: high shelves and anything I dropped on the floor, the building on the other side of a gravel path. The gym felt big too. The weights that a training instructor handed me felt large, even though they were just a kilogram. I looked at the other wounded from my gym mat, all at different stages of their recovery, lifting coloured balls or bars, or stretching bands. Coming to terms with a life-changing injury is an interior job, but I would quickly get to know these people. I wasn't alone – we were a community. We were unexpected survivors, young and fit and with a potential for recovery that made a mockery of 'likely outcomes' based on out-of-date studies.

One afternoon, a few weeks after I arrived, I saw a double amputee with similar injures to mine walk past in the corridor. I scrabbled from my bed into my wheelchair and propelled myself to the door to watch him. He didn't have sticks and he ignored the lifts and disappeared into the stairwell. I'd get to know him as a shy Marine who needed another sort of inspiration, but back then, when I was the new boy, it was like seeing a Year 11 student on your first day at school – impossibly confident and intimidating – and being unable to imagine ever being that grown-up.

I was starting a life I had little understanding of. My idea of the amputee, like many people's, was stereotyped and polarised to extremes. At one end, I saw the homeless beggar on a street corner with one trouser leg empty; and at the other, I saw the superhuman sprinting the corner of the running track on a carbon-fibre blade or dancing once again on a microprocessor knee – injury overcome and living a normal life. Maybe even a special life.

During those first weeks in hospital, when I allowed myself to think of the future, I found it hard to imagine walking again; the body I looked down on just seemed too pathetic and incomplete. I accepted this – survival seemed enough. Life in a wheelchair felt most likely, maybe with prosthetic legs for special occasions, tottering forward on sticks. I could still find hope in that. But it didn't take long to find uplifting stories: they were in magazines and on the internet, and walking past me in the rehab centre. It made me impatient. *How soon can I walk, when will I run, when will I be myself again?*

The possibilities were made all the more exhilarating as we seemed to approach a technological horizon where the dreams of science fiction lay within reach, where artificial body parts had the potential to outperform what they replaced. I watched para-athletes compete alongside the able-bodied, bionic hands move with the power of thought, and the paralysed walk again with the robotic assistance of exo-skeletons – technology and flesh combining to overcome human frailty and loss.

Then, ten weeks after losing my legs, I stood on prosthetics for the first time. I lowered my weight into the sockets and it felt as if a vice had closed around my stumps. The wounds were still raw from surgery, and everything that had been healing for the last eight weeks crushed together. An electrical storm of impossible sensations fired – a drawing pin trodden on, a toenail pliered off, salt rubbed into a skinless foot. It was a moment of breathtaking pain and weirdness. It was also the sudden vertigo of unbending my stiff back and being 6 feet tall, after so long in beds and wheelchairs. But standing between the parallel bars in the limb-fitting room, arms quivering as they supported my weight, I looked up and saw my body completed by prosthetics for the first time and smiled.

My physio, Kate, and prosthetist, Mark, who had suddenly become just about the most important people in my life, were watching, poised close enough to catch me if I fell.

'How does it feel?'

They wanted to know if they could improve the set-up of

the legs, but they were also smiling – they had an idea of how standing, after losing your legs, might make a person feel.

I sat back down. It was painful and the sensation was alien and hard to describe. I made an attempt – and said something I'd never said before: 'It feels like I'm standing on the end of the bone.'

They told me to remove the leg and gave me a small white sock to roll over the end of the stump. 'You're dropping in. You'll be losing volume all the time as the inflammation of injury goes down. A sock or two might make the fit better until we cast you again.'

I pressed the leg back on and stood up from the wheelchair. 'Try taking a step,' Kate said.

I took that first step, a little step that made the prosthetic move out in front of me and touch back to the carpet with a spike of pain. It was different from anything I had experienced before: the thoughts that used to result in *step* no longer had the outcome I expected, and the collection of components suspended off my stumps wouldn't respond. The prosthetics felt frustratingly dead hanging there, and I could only animate them with exaggerated movements from my hips. I hissed out my frustration: *Move, go on, step.* But there was only a void where once there had been sprightly, quick flesh. My brain still thought I had legs, yet my nerves were signalling to muscles that were severed in my stumps. The lack of response to the commands returned as pain, unstable steps scuffing the floor and the prosthetic knee suddenly giving way.

This was exactly what happened as I tried a few more steps.

Right after Kate had said *well done*, a gathering familiarity made me overconfident, and the attempt to brace my real knee (which had been surgically amputated and incinerated) caused the mechanical replacement to buckle. Falling would become normal in the months ahead, but that first one was terribly painful. Reflex made me jerk out a foot to stop the fall. A real leg – the old me – would easily have stopped this stumble. Instead the ghost of my foot disappeared through the floor and the prosthetic sheered away from my stump, scraping the raw wounds against the inside of the socket. The pain was not only physical; it was the pain of my loss made real.

Mark and Kate helped me back into the chair. White and sweating, I took the prosthetics off. A wound on the end of my stump had reopened and dripped perfect spheres of blood onto the carpet. Kate handed me a paper towel to blot it. Enough for today. I slid off the treatment table into my wheelchair and rolled out through the automatic doors. Up on the ward, a nurse cleaned and dressed the wound and I went to collapse on my bed, stumps buzzing with the release of pressure and phantoms ghosting from old nerve endings.

Each time I returned to the limb-fitting room, wheeling my chair over to pick out my legs from among scores of others propped along the wall, I found it easier. For all the newness of the prosthetics, there was familiarity. They inhabited the space in which my brain expected my body to be, making me upright again, letting gravity act on me in a way that felt normal. What was so odd was how natural it looked.

Suddenly there were shoes down there, where I was used to seeing them. And walking up and down the parallel bars became a few tentative steps across the limb-fitting room with crutches; walking with two sticks along the corridors of the rehab centre became wind-blown walks down gravel paths in the garden. Each time it was a little easier. And the strangeness of it faded as I improved.

Although I never wanted this body, I started to enjoy myself. The human ability to adapt made learning to walk for a second time exciting – for all the pain and weirdness, there was also joy. The physical challenge and gradual improvement were an antidote to the trauma. I had a purpose to focus on: walking again after thinking you might never be able to was close to touching magic. A future with a different body now seemed hopeful, a life of new experiences that I might not otherwise have had, and I knew I could be independent again.

And then, after a year of rehab, I was discharged, sent back out into the world with a new body.

While rehab had been uplifting, I know now, learning to live in the everyday of the real world was a different challenge. Like winning the gold medal or achieving a lifetime goal, there was a moment of depression when I was finally discharged. The intense purpose of physical rehabilitation was gone. *What now?* It was deflating. It was also the realisation I wasn't going to get any better. That this disability was for keeps. That I would have to live the rest of my life managing my body, and the tech that assisted it.

*

I quickly found other goals. I retrained, I started a new job, I married, I had children. I was happy. Ten years have passed and now I can carry my kids on my shoulders, commute to work, walk beside my partner holding hands – being an amputee feels normal. I also look back over those ten years and see how much I have changed. I have a new body and a different identity. I also see how dependent I am on the technologies that help me; without them I'm not the same.

But a decade after discharge from the rehab centre, I need to check again if I've left any stone unturned: can I find new and better solutions? Can I improve my relationship with the technology I use? And perhaps most important – I am so utterly different from the person I would have been, had I not stepped on a bomb – can I understand the impact those technologies have on the person I've become? Ten years after my unexpected survival I have entered a kind of adolescent introspection: I need to define who I am, how I fit in, to measure myself against society and my peers.

I look around me and see that we are all experiencing a deepening and more intimate relationship with technology. We are all, in some ways, plugged in, and this changes our bodies and our brains. At the same time, distinctions between being disabled and able-bodied seem to be blurring – so many more of us support our imperfect biology with technology. It feels like we could be moving away from *disabled* as a useful description. It doesn't quite sum up what's at stake, especially as there seems to be a possibility, not far in the future, when technology might make me more capable than an able-bodied person – or at least in the strange position where I am as capable when I have my prosthetics on, and far more dependent when I don't. The paradox already exists: to the casual glance, I walk as fast and upright as anyone else in the street; the cost and effort are hidden, and outwardly it appears that technology has fixed me.

I've always felt uncomfortable with *disabled*; I know it is a legal category that protects me and I benefit from, but I've pushed back. I don't want to be labelled. Nearly all the technologies I've learnt to use, and so much of the last ten years, have been about overcoming my physical impairments – what would it be to live in a world where disability is normal and truly accepted, where the attitudes and structures of a society that create the disabled disappear?

I am wary of new categories, but I look around for anything that might better describe how I feel. *Cyborg* and *bionic* carry too much baggage; they conjure too many fictions and unrealistic expectations. *Differently abled* seems a word game. So I have started to think of myself as a *hybrid human*. It is a label just for me, not one I would impose on others. Hybrid bikes and hybrid cars, hybrid working – it is on-trend; it isn't perfect, but 'a combination of two different elements' seems to fit. And I like that *human* is part of it – human, more than anything, is how I want to feel, and it disappears from cyborg, bionic and disabled. This hybrid is a fusion, an amalgam, a confluence of things – of pig's-heart hybridity, of robotics and AI (artificial intelligence), of genetic engineering and new kinds of interfacing. It feels like a better way of describing my experiences and is somehow less loaded.

What if I was no longer a disabled person, but a hybrid human?

What follows is a journey to find stories at the frontiers of man and machine, and to ask what it means to become dependent on medical technology; how the everyday realities of replacing and enhancing the body can change a person; and what the people at the extreme fringes – where monsters and cyborgs lurk, where technology has the capacity to dismantle identity and the traditional expectations of society – tell us about who we are.

Metal Ghosts

I'm somewhere deep in the enormous London-museums repository in a long, low, windowless room. At the far end a broken strip light flashes among the shelving and in this corner, pushed together like wheelie-bins at the back of an office block, are the iron lungs – cream-coloured coffin-shaped boxes that could be props from the deep-space stasis scene of an old sci-fi movie. One of these metal tanks has a head sticking out of it, collared by a rubber seal and staring at the ceiling; and when I look through the inspection windows – portholes for nurse and carer observations – I see the rest of him, frozen in time inside the machine. He'd have been wheeled back in here after his last outing on public display, for the long sleep among all the other objects. There's something about the mannequin, trapped motionless in this box and locked away in the dark, that stands as a shrine to the people this machine served.

*

I'm one of the last people to be allowed to visit Blythe House, home of three national museum collections. The 5½-acre site is being sold – it will be flats and a hotel – and the 300,000 objects of the Science Museum collection are being sorted, categorised and packed, ready for transport to a more accessible home at an old RAF base in Wiltshire.

Waiting in the reception area was slightly unsettling. It's very like a prison and I had a weird sense of déjà vu. I half-expected a buzzer to sound for visiting hour. Stewart, the curator who has agreed to show me around, explained as we walked into the building that I have probably seen it before – it's often used as a location for TV and film. It was the headquarters of the Post Office Savings Bank before it was handed over to the museums in the late 1970s, and Stewart points out a few traces of the building's old purpose as we descend the staircase that takes us to the rooms holding objects from the history of medicine.

In the late 1920s the engineer and industrial hygienist Philip Drinker used the negative-pressure principle to design *An Apparatus for the Prolonged Administration of Artificial Respiration* – the first iron lung. There's a photograph from the time of Drinker standing over a small box. Look closer and you see a cat's head at the end of a box, gums drawn back and teeth bared. He had triggered respiratory arrest by paralysing a series of cats with South American dart poison and (after a few dozen failures) proved his concept by pumping his machine and keeping the animal alive until the poison wore off. The machine that Drinker went on to patent enclosed the human body inside a metal tank; the head protruded through a rubber collar at one end, and at the other was a pump that raised and lowered the pressure inside the airtight chamber. Because the patient's head was outside the machine, sealed off with a rubber collar, lowering the pressure inside the tank caused the patient's chest to expand and air rushed in through the mouth. Raising the pressure in the tank compressed the ribs and the air was exhaled.

The iron lung quickly became associated with polio. The virus peaked in parts of Europe and North America in the 1940s and 1950s – coming in the summer, spreading fast and leaving before winter, it paralysed or killed half a million people worldwide each year. There was no way of keeping track of how many were infected – 85 per cent had no symptoms. The symptoms of those who did were mostly mild: sore throats, fever, headaches and vomiting. Yet the disease caused mass anxiety. Parents dreaded the summer outbreaks. Birthday-party invites were ignored and playgrounds were left deserted. When a child developed a fever, parents prayed it was just a common cold. If children lost mobility in their limbs, there was a helpless wait to see if they would be one of the unlucky 0.5 per cent who had paralytic anterior poliomyelitis, which could have devastating effects on the central nervous system. In severe cases, paralysis spread to the intercostal muscles between the ribs and patients were unable to breathe and died.

By the middle of the century, hundreds of iron lungs lined hospital wards, many of them used by children. The noise of the pumps whooshed, nurses checked the pressure and breathing rate, adjusted the tilt of beds and turned the pages of books suspended on stands above the children's heads. Most patients spent a few weeks in the iron lungs, clutching a cuddly toy or doll in a splinted arm, before they recovered enough to breathe on their own. In a small fraction of cases, paralysis was more permanent and they became dependent on the huge, unwieldy apparatus. An iron lung would be installed in the corner of the living room, a massive presence that took over family life. Some might manage outside the iron lungs for a few hours, heading to school or college or work, then returned for respite. Some would sleep in them; and a few, where the paralysis was total, would spend their whole lives cocooned. By the early 1960s the vaccine, which had taken fifty years to develop, had largely eradicated polio, but the iron lungs continued to breathe for an unlucky few.

I touch one of the cold, hard lungs. It must have needed three or four people to move it. I imagine myself locked into a tank by nurses. Wanting to itch my nose. The rubber cuff sucking and puffing around my chin. The claustrophobia. Yet the panic of struggling to breathe was so dreadful that many polio victims described the relief of being clamped inside the machine for the first time and letting it breathe for them. I think of the moment I first joined myself to prosthetics in the limb-fitting room – the fear, the relief, the pain, the weirdness of it. These iron lungs make for even stranger hybrids.

Nestled among some of the adult-sized machines is a barrel-shaped lung just big enough for a baby or small child.

26

Stewart fidgets behind me. 'Some of the team won't work alone in this room,' he says.

I turn to look down through the shelves. 'You?'

'I prefer the door open, but I'm okay.'

'I've never seen an iron lung before,' I say. 'If anything should be shown to the anti-vaxxers, it's these.'

Stewart turns off the lights and locks the door. We cross the corridor and he opens another room. Rows of cabinets and vitrines and shelves filled with medical objects from the past. I stop in front of the spectacles: lorgnettes, pince-nez and more modern bifocals. Next to them are the glass eyes, staring this way and that from a silk-cushion tray. I want to touch one, but don't. Stewart is loitering. These are from the late nineteenth century. There is a variety of iris colours. The fine-painted pink capillaries beneath the sheen of glass are uncanny.

Other than preventing scarring in the socket and keeping muscles working by giving the eyelid something to blink over, a glass eye has no function. It is a cosmetic prosthesis. But to the wearer, it must have meant much more. People stare at an empty socket or eye-patch. With a glass eye in, life is easier. Fewer stares, and more likely to get that job, more likely to woo that partner. I wonder how the owner would have felt, lifting their upper eyelid and slipping in the lozenge-shaped prosthetic and seeing two eyes reflected back at them?

Many of the objects in this room might look like tools from a torture chamber, but this is the stuff of healing, diagnosis, pain relief – of making the body complete again. Keeping humans being, and feeling, human.

Stewart flicks the light off and locks the door behind us. He has one more room to show me.

I'm a husband. A father. A son. A brother. A friend. I'm British. A Londoner. I am a straight, white male. Six foot (I used to be 6 feet 2 inches) and of average build. I like watching sport, but don't have any teams (other than England when they're doing well). I'm dyslexic. I'm a veteran and have been to war, but don't think about it much. Once I nearly died, et cetera.

This list shuffles around, depending on where I am, who I'm with and how I feel. And somewhere into that list I have to fit these new identities: *amputee*, *disabled*, *medicalised*. When I was in A&E a few years ago, waiting with pain and anxiety for antibiotics to treat an infection in my stump, these terms were all very near the top of the list. Last week, catching up with close friends, not on it at all.

If that's my identity – my sense of self – then what makes up the physical *me* (the disparate, interconnected and entangled stuff of a body) is just as slippery. The interplay of genes and environment again provides the underlying blueprint, influenced by everything from the final fag my mother had before she realised she was pregnant, to what I had for breakfast this morning – all shape the body I have at this moment. Chemically speaking, I'm made up of around sixty different elements. Biologically, I am 65 per cent water. And around 3 per cent of my mass isn't human at all; it is 10,000 or so completely separate types of organisms: bacteria, fungi, protozoa and viruses that live all over me, mostly in my gut. Although this microbiome represents a small percentage of me, there are 100 trillion of them. They outnumber my own cells by ten to one.

But my body – the one I identify with and embody – is atypical. Of my 68 kg, 60 are flesh-and-bone wetware and 8 kg are prosthetic hardware. (That's almost all legs, plus a few milligrams for my contact lenses.) This makes *me* 12 per cent machine. What does that mean for how human I am? If being able to walk alongside Stewart, at a similar height, and look him in the eye is some measure of how human I feel, then that 12 per cent is critical to my humanity.

If you removed my legs and left me on the floor, the list would shuffle again. *I am shameful, I am vulnerable, I am less of a person* would rise quickly to the top – it would be very similar to stripping me naked in public. And I'd have to shove along the floor on my bottom, swinging my body between my arms to keep up with Stewart. Perhaps that's why I am so moved by these objects; I have some idea of what they meant to the people who used them.

In 2008 there was a storm over Jackson, Tennessee. Trees fell on the power lines near Dianna Odell's house and there was a power cut. She had spent sixty years in an iron lung. That night the emergency generator failed. Her family desperately worked the manual pump, but she died. She'd caught polio when she was three, the paralysis spread and she couldn't breathe. She lived in her front room, cocooned in a technology that noisily breathed a lifetime for her. She passed all her schooling, received a degree and wrote a book, all from the confines of her iron lung. She said in a press interview,

'I've had a very good life, filled with love and family and faith. You can make life good or you can make it bad.'

There are others: the man who ran a law practice from his iron lung; the artists who clenched brushes in their mouths to paint; the woman who worked behind a desk in her local bank branch, returning every evening to be shut in an iron lung by her mother.

When I'm not at home, my eighteen-month-old son sometimes takes one of my liners (the silicone socks I wear over my stumps to hold on my prosthetics) from the clothes dryer, walks into the kitchen, looks up at my partner and says, 'Da-da' and hugs it. Maybe that's what I can feel here – the trace of people so dependent on this equipment that it became part of them.

Carbon-fibre feet and pylons, titanium adaptors, grub screws, valves, composite sockets and silicone liners – 8 kg of technology allowing me to walk down the corridor with Stewart. I have my biological left knee (I'm amputated a few inches lower on that side); it's the technology that stands in for my right knee that I am most reliant on. As I walk, it adjusts to my gait. If I stumble or stop or take a step backwards, it will adapt and prevent me falling. It knows, for example, that I have now started to descend steps, following Stewart, and is lowering my weight predictably.

I've tried a few of these bionic knees over the years, and this seems to be the best all-rounder. It's a Genium X3, the latest generation of microprocessor knees produced by the German company Ottobock. The marketing material says it's *incomparably close to nature*, and goes on to list features that read like the technical spec of the latest German car: Dynamic Stability Control, sensing to 1/100th of a second the change from stance phase to swing phase; the Internal Motion Unit, with its gyroscope and acceleration-sensors measuring where the leg is in time and space; the Intelligent AXON tube adaptor, gauging ankle movement and vertical force; the Bluetooth function to link to an app on your phone and change settings and modes – the knee can cycle, run, play golf, it can even be set up for ice skating; the hydraulic unit with its two control valves; the knee-moment sensor; the battery with five days of life, charged through an induction plate; and all of it housed in a carbon-fibre frame and *extra-robust* polyurethane protective cover. (The German car comparison goes further: 'on the road' with the six-year warranty, this 1.7-kg unit costs around £70,000.)

At the centre of all of these components, and making the decisions, is the microprocessor. This little chip receives information from all those sensors and feeds it into a control algorithm, which, depending on the countless variables and permutations, decides what to do next. If it's quiet – say, at the end of the day, when I slump down on my bed to take my legs off – I can just about hear the microprocessor *thinking*, crackling away. Sometimes I bring it close to my ear to listen. It makes a sort of electric squelching noise as it controls the valves in the hydraulic cylinder that change the resistance of the knee.

It really depends on how you define artificial intelligence, but if this knee is a device that *perceives its environment and*

31

takes actions that maximise its chance of successfully achieving its goals, then perhaps I do have another brain down in my knee, and I should add it to the list of stuff that makes up who I am. Yes, it's dwarfed by the computational power of my brain (the control I am exerting through my hips over the whole system – that is, me and the prosthetics – is far more nuanced than anything it is doing), but there's no way I could keep talking to Stewart while descending the stairs without the extra control this little *brain* is giving me; I'd be thinking too hard of not falling. I've delegated some of the cognitive load of walking to a second brain, and it's an important part of me and the way I experience the world.

Stewart unlocks the final room. It is bigger than the other rooms he's shown me – about the size of a squash court. Metal shelves divide it in rows, like the aisles in a library. On the shelves are the Science Museum's collection of orthotics and prosthetics. The soles of scores of feet face me on the right – wooden, leather, plastic – lined along the racks. It's like looking at the end of a very crowded four-storey bunk bed. They are legs from the last hundred years or so. The years after the world wars are particular well represented. I walk the aisles, trying to take it all in. There's a leg that looks as if it's from a suit of medieval armour, riveted and shining silver, but the label says it's *aluminium with central knee control, and wood foot with toe and ankle joint – made c.1920*. Then a dozen or so peg-legs and a pile of crutches.

We often pine for a better time, feeling nostalgia for the past. But the advances we've made in the last few decades make everything here look archaic; and, standing in front of these shelves, I feel glad I'm disabled now, with my trusty microprocessor knee holding me up. There is no better time to be an amputee, and probably no better time to be in any way ill or disabled.

The next aisle has orthotic shoes and an unnerving collection of ceramic children's dolls. One is in a scoliosis brace and splints, for 'drop foot', complete with little teddy and lying in a little bed. The label says it dates from *1930 to 1950; probably used to demonstrate to child patients a convalescent stage in their prospective orthopaedic treatment*. Next to it is a little model of an iron lung.

In the middle aisle I stop in front of a shelf with a two-arm prosthetic contraption. It's worn girdled over the shoulders and looks desperately uncomfortable. The label says it is powered by *a motor-driven car assembly*. The arms are a mechanical automata of cylinders, wires and pulleys – sized for about a five-year-old. There's another child's upper-limb prosthetic with a split hook next to it, *powered by compressed air*. This is the cabinet of the relics of the late 1950s and early 1960s thalidomide disaster, which led to thousands of congenital malformations in children – the attempts of prosthetic and bio-engineering departments to right the wrongs of the scandal. But most of the patients with limb difference caused by the drug still found life easier without these contraptions – the feel and function of their residual limbs couldn't be improved on by medical science, and they sit here, failed experiments.

In the last aisle I crouch to look at a small peg-leg. It's

made from a turned wooden chair leg attached to a socket and hinged knee brace. The label says it was *made in a shipyard by a father for his three-year-old son in 1903.*

It's hard to get a grip on how much time this collection spans. Stewart is not sure. He puts on blue plastic handling gloves and picks up an upper-limb prosthetic. It's made of dulled iron-plate work and has some sort of simple ratchet-and-gear system in place of tendons and muscles. The fingers are fashioned from more plates of sheet metal formed into a grip. Stewart holds out the label for me to read: *Artificial right hand, iron, ?, owned by Götz von Berlichingen, 16th century.* The question mark is ominous, and Stewart says as much.

'Take this with a pinch of salt. We had an expert look at it, and he thought it was probably a more modern replica – maybe from some drawings of Götz's hands.'

Götz von Berlichingen (1480–1562), or *Götz of the Iron Hand*, was a German imperial knight, mercenary and poet. His hand was taken clean off by a cannonball during a siege. He had a metal prosthetic made for him by an armourer, and continued to soldier. It's said he could still grip a sword or reins, even a quill pen, with the iron fingers. He went on to become famous – a German Robin Hood – and, when not employed by kings or emperors, wreaked havoc, kidnapping for ransom and attacking convoys of merchants. His hand was a technological wonder and became a symbol of national inventiveness.

It's true, there's something not quite right about the limb that Stewart holds; the way the thing is put together doesn't feel authentic, as if an archivist had added Götz's name in a rush of vain hope, one Friday afternoon years ago, and had never got back to check it.

Limb loss and difference have been around as long as humans have suffered disease, trauma, punishment and congenital anomalies. For most of history, without the antibiotics and wound treatment of modern medicine, the first challenge, particularly with traumatic amputation, would have been survival – in some cultures, societal rejection for not being 'whole' would have made it harder – but it wouldn't have been impossible. (Mammals that recover from traumatic amputation have been recorded thriving; African wild dogs have been seen supporting a three-legged member of the pack by giving it a turn at the carcass.)

Every once in a while archaeology opens a window on how disability might have been treated in the past. The skeleton of a man who lived around 4900–4700 BC was excavated near Paris. He was missing his elbow and lower arm. Analysis of his humerus showed that, after traumatic injury and some surgery, layers of new bone had grown, proving he had healed and lived for months, if not years, after his limb loss. It's one of our earliest examples of a surgical amputation and hints at remarkable prehistoric medical skills. And we know this man was not shunned by his Neolithic farming community. Buried around him were sepulchral goods of a quality rarely seen in similar digs: a flint pick, a polished stone axe, the skeleton of a votive animal. All suggesting that, despite his deformity (or maybe because of it), this man had been a significant member of the team. It seems that community care and solidarity were important in this area of northern Europe nearly 7,000 years ago.

Time has disintegrated almost all evidence of the aids for rehab and mobility made by ancient cultures. Yet it's easy to imagine our ancestors were resourceful enough, repurposing and fashioning rudimentary wooden splints or simple prosthetic attachments from what they had around them. We have only a few examples: 2,200 years ago a man from Turfan in China lost the ability to walk. He had tuberculosis and his knee fused in a 135-degree bend. He used a wooden prosthesis attached to his deformed leg, with a stabilising thigh corset. The leg had been wrapped in ox horn, reinforced by goat horn and tipped with horse hoof for grip, and the whole thing was worn, from years of use. His skeleton, along with the prosthetic, was uncovered in 2008.

The oldest-known working prosthetic was discovered in the necropolis of Thebes in Egypt, attached to the right foot of the mummified Tabaketenmut, the daughter of a priest, and dated to 950–710 BC. She was perhaps one of the first hybrids. Her big toe was made in three sections, of wood and leather, and was stitched together. It was carved to be lifelike, with a curved nailbed. The human big toe carries 40 per cent of our body weight and propels us forward – making a mechanical replacement a difficult problem to solve. But this toe's hinges, the wear and tear and the holes for attachment all suggested that this almost 3,000-year-old prosthetic was functional as well as decorative.

*

'This would have been our oldest,' Stewart says, pointing at the bottom shelf. 'But the original was destroyed in 1941, during the Blitz. Luckily, someone made this replica around the turn of the century.'

It looks like a smooth lichen-covered log at first, but then I see the way it shapes from ankle to calf. It's made of bronze and wood, with fastenings where leather straps would have attached it to the thigh or waist. I read the label: *Copy of a Roman artificial leg, c.1910. Original had been in the Royal College of Surgeons and had been in a grave at Capua, c.300 BC.* I see the burnished bronze leg glinting beneath a toga, walking along a street 2,300 years ago. Surely an unusual sight – the Roman equivalent of my microprocessor knee. I have a sudden confusion of imagery: the leg being made in the foundry of a hot Roman market town; a soldier who walks after injury in a war of swords and shields; the leg buried with him, surviving as he decays; dug up to become a wonder of antiquity, an artefact of display cases and academic papers; and then destroyed in another war of planes and bombs.

An epilogue flickers, almost too hard to conjure: my own prosthetic knee lying on a museum shelf 2,000 years in the future; my body long dust, this technology being the last trace of me. I can't picture the other medical devices catalogued around it, or what sort of a body the person who is looking at my leg might have.*

---

*A curator at the Wellcome Collection asked me if I could loan a leg for a new permanent exhibition, *Being Human*. I gave them a C-Leg, the first microprocessor knee I was prescribed. At the gallery opening

*

The dominant Western approach of philosophy and science has been to understand the human body as secondary to the mind. *I think, therefore I am.* In the information age, our brain becomes a biological computational device creating our mind in the same way a computer runs a program. But there's something about having a body. Without it, our thoughts and actions would be empty. Embodied cognition is the idea that the way we think and perceive the world is deeply dependent on having a body. Our movements and sensations become essential in underpinning our thoughts – and mind and body cannot be untangled in any way that makes sense.

When we have a body that is dependent on med-tech to function properly, we are aware of this embodiment in different ways. Back during my rehab, I'd been surprised at how quickly the assistive technologies I used felt part of me. First the wheelchair – knowing, for instance, exactly how wide it was, and how to balance it on its rear wheels – and then the prosthetics and the way I incorporated them into my body. The repeated action of intense learning strengthened my neural pathways and physically changed my brain. This principle of neuroplasticity – which lets the brain continually adapt to changes in the environment – allowed me to *rewire* in response to the new experience of disability.

– champagne and speeches – it was strange looking at one of my own legs inside the glass box, and I left before anyone would notice that a part of me was on display. Perhaps it will be preserved in museum storage long after I am dead.

We used to think the adult brain had limited capacity for change, that we were stuck with the brains our childhoods left us. But during the first two decades of the twenty-first century (and in large part because of the advent of fMRI scanners, which can show us in real time different areas of the brain lighting up as we think) our understanding has shifted. The adult brain is far more plastic than we once believed and can respond to learning and experience by physically changing itself, even into old age.

We can extend ourselves to incorporate a tool instinctively, almost as soon as we've picked it up. We are a tool-using species, capable of learning to operate highly elaborate instruments. We can 'feel' how big a car is and thread it through the narrowest of gaps. Our brain deals with the problem for us by creating a representation of the large box-like space on wheels that we are inside. (That's also why we sometimes duck when driving under a low barrier going into a car park, even though only the car is any danger of being hit; we have become the vehicle – we embody it.)

I was able to merge myself with the prosthetics so they felt part of me. For anyone who acquires a disability, the knowledge that we can combine our bodies with technology so intuitively should be uplifting. Embodiment is the universal compatibility of the hybrid human – we are set up for *plug and play*.

Time is almost up, and Stewart shows me a tray of upper-limb attachments as we move towards the exit.

'These might be of interest.'

At first glance it is just a tray of vintage household tools, but there are more industrial-looking implements in there too. I read through the various labels: there's a brush, a driving ball, a chisel-holder, a hammer, an adjustable clamp and guide for a screwdriver, a spiked device for holding food when cutting it up. They all have an attachment that clicks into the socket of a forearm prosthetic to replace an amputated hand – it could be a drawer in Captain Hook's dressing table. Many were made after the Second World War, for getting the boys back to work. I close my hand into a fist over the tray and imagine slamming it down onto a nail. I replace my hand with a hammer and repeat the action; it's easy to imagine. The nail is driven down with a clunk. It would be easy to embody a hammer in place of my hand – as easy as holding a hammer, perhaps easier.

We all possess a body and we are all, in some sense, possessed by that body. For the most part we don't even register that our body is 'on', other than perhaps in those first few moments of the day when we struggle to boot-up in the morning; or if we are stressing ourselves with exercise or we are ill or hurt, then we are aware of it. But mostly we forget about our body. Yet I can't help feeling that those whose bodies are dependent on technology are far more likely to feel that possession and to notice their own embodiment. It's harder to forget your body when you're disabled.

I take a last look at the hammer prosthetic, then push the tray of tools away. How would replacing someone's hand with a hammer make them think differently about themselves? A body part that is for feeling, for hugging, for connecting with

40

others, changed into an implement that is for hitting. The body made tool.

Stewart shows me out of the building with a quick hand-shake. All I can think about is the people who used those objects, now stored in the dark. I can hear it: the child calling for her mother as she was strapped into the upper-limb prosthetics; the groan of the ninety-year-old man pulling on his leg; the tears of the veteran when no one was looking – a hammer instead of a hand, his body repurposed so he could work in a factory; the mother kissing her daughter goodnight as she shut her into an iron lung. Every one of those objects was embodied by a person. They repaired, remade and sustained them, but they also changed their bodies and their identities.

As I walk, I can see my prosthetics stepping into my peripheral vision. They are not human. But when I am without them, I feel less alive.

Interfacing

I knew Jack from the earliest days of my rehabilitation. The night after seeing him I try to ready for bed, using one arm. I hold my left forearm tight across my stomach and pretend it has also been blown off. To remove my prosthetic legs I press a small valve, breaking the vacuum between liner-seals and socket wall, and pull my stumps out. With both hands this is easy, a single motion, I can press and pull, but with one arm it takes more than a minute. Then I try to peel off my liners – they are made of rubbery silicone and I can't get the purchase I need with one hand. There's a small split and I rip it open in my growing frustration. (These liners cost £400 and need looking after. I'll have to order another through the NHS.)

When I'm free of all the prosthetic clobber, I slide across the bathroom floor with one arm and bum-shuffle into the middle of the shower tray, then stretch up for the controls but can't reach. Normally I push up off the floor with one hand and turn with the other. There's a huge temptation to use both arms now. I manage to reach by hopping my body up and knocking the lever. This is painful and I can't get away from the cold water, and I gasp and sit under it as I wait for the heat. Everything is difficult with only one arm. I can't pour the shower gel into my other hand, so end up squirting it straight onto me, and I find I have to wash my arm and hand on my chest and the suds go everywhere and in my eyes.

In the shorthand slang of the veteran amputee community, I'm known as a 'double', and Jack is a 'triple'. Just designators for the number of missing limbs – and you might qualify it by adding 'above' or 'below' to indicate the height of the amputation in relation to the knee. (So when describing someone whose name you've forgotten, you might say, 'You know, that double-above who was always smoking during rehab and has loads of tats and piercings now and works as a mechanic – yeah that's the one.') About seven years ago a new phrase entered injured-veteran slang. At a charity fundraiser or on social media, you'd hear something like this:

'Mate, he's going for osseo.'

'*Osseo?*'

'Osseointegration. He's gonna fly to Australia and they drill out your thigh bone and stick a rod out of your stump.'

'Shit, mate – sounds drastic.'

'Yeah, but no sockets. They reckon people who have it can walk around all day, no dramas.'

'Australia?'

'Can't get it on the NHS, he's going to have to fork out for it himself. No sockets, though … How good would that be?'

Osseus is Latin, meaning bone – and *integrates*, to make whole. In this case, the integration is the bone of the residual limb, the femur, bonding with a titanium implant that extends out through the skin. A prosthetic leg is then attached directly to this. It's sometimes called 'direct skeletal fixation', which better describes its benefits (and some of its risks). By attaching directly to the skeleton, many of the challenges of how to join the body to a prosthetic are overcome – there is no need for any straps, liners or sockets. For some amputees, where the traditional rehab route of sockets hasn't worked, osseointegration is life-changing.

But it wasn't offered in the UK in the early days – there were just too many risks. Because the implant sticks out through the skin, creating an open hole (or stoma), there is a permanent breach in the skin through which infection can enter; and to receive the implant, the bone has to be cored out during surgery, increasing the chance of fractures and deep bone infection. It's also irreversible. The NHS wouldn't sanction the procedure and said it wouldn't be responsible for the complications or long-term treatment of those who had gone private.

But then a couple of soldiers took action. They'd never really been able to get the most out of their prosthetics – they'd tried every socket and every liner, switched prosthetists and limb-fitting centres – and had reached the end of the road. Their stumps were just too short. They lay on their beds, glancing over at their legs propped in the corner, and scoured the Web for anything that might help. They wouldn't accept that there wasn't a solution – particularly as they watched the friends they'd rehabbed alongside walking their way back to normal lives. They found a website with an amputee without sockets – her prosthetic seemed to float below naked stumps. They saved all the articles and videos and took them to their next appointments, to make the case to their doctors. They faced sucking-through-teeth cautiousness: too expensive, too experimental, too many complications and too high a risk of infection.

But instead of accepting their circumstances, and with nothing to lose (what's the worst that could happen? they were stuck in wheelchairs and would only end up back there if it all went wrong), the soldiers saved money, begged funds from charities, sat on breakfast TV couches to put a bit of public pressure on the policymakers, then travelled to Australia to have osseointegration done privately. They returned walking further than they had since injury, often far further than other amputees who had previously been way 'ahead' of them. Their gait was better, they were getting fitter, were happier and were almost pain-free.

And you started to hear 'osseo' more frequently among the community. (*'He's a double osseo now. Yeah, he's crushing it. I'm going to go for it.'*)

This was a problem for the system. These amputees had returned with far better outcomes than expected, and word was getting around – lots of *doubles* and *triples* wanted osseo now. The few early adopters really were *crushing it*, but there was still the potential for complex needs if it went wrong.

No one knew how long the implants would last, or what would happen if they failed. Whatever the policy, the NHS would have to look after those who'd gone private if they had complications. A torn liner is easy to fix; the new kinds of hybridity that osseo promised meant far more complicated care.

In a surprisingly agile response, two surgeons who had spent the last decade of their careers patching up soldiers decided to bring the procedure to the UK. Despite some reservations, they saw how it could be game-changing for some – and felt strongly they had to stop people heading abroad to have it done privately, where anything could go wrong. They received ethical sign-off, and the money to trial the procedure on a dozen or so soldiers from Libor funds (remaking soldiers being the perfect use of the penance payments of the bankers).

Jack had turned up at the rehab centre a few months after me, one of the very young soldiers who appeared, having lost three limbs in a blast. I liked Jack. Despite his injuries, he was always smiling and joking and was a genuinely warm presence on the wards – to think of him was to see him smiling and laughing from his electric wheelchair. A few times we shared the same bay: you could tell he was finding life tough (which eighteen-year-old wouldn't, after losing three limbs). He is a big man – good for soldiering, bad for amputee-ing – and had been left with short stumps. None of the sockets that were made for him seemed to work, and while most of us got up on our new legs and made progress, Jack struggled. I liked him for not hiding his frustration in that macho way some soldiers did (a sort of strange mash-up of gangsta-tough-guy and Brit stiff-upper-lip). He was open with the rest of us about how crap it all was, but still managed to be upbeat. So when I heard he'd put himself forward for the trial and was now a *triple osseo*, I gave him a call and we made a plan. I wanted to see how he was getting on. I also wanted to know what it was like.

We sat at Jack's kitchen table. He had his short stubbie legs on,* and the chin of his low-slung British bulldog, Bruce, resting in his lap.

After we talked about the new house renovation he'd just finished, I asked if he had known straight away he wanted the procedure.

'I was pretty good friends with Michael Swain,' he said, 'the first guy to have it done. He'd researched it and found the team in Australia. We were all in the day-room in rehab and Michael told us what he planned, and were like "Fucking sounds extreme" and wished him the best – "You're the guinea pig," we told him. And then I saw how well it worked out for him. As soon as they started the trial, I put my hand up.'

'Were you anxious?'

'Of course, yeah. But I had more anxiety when I was on sockets. They never fitted. I could never walk properly – that outweighed whatever issues I might have with osseo. I was in

---

*Stubbies are short prosthetics that don't bend, often used during rehab before progressing to full-length legs, which tend to have a mechanical or microprocessor knee.

my wheelchair most of the day; I wore my stubbies for exercise, but that was about it. I saw the other guys, like Michael, getting on really well. I thought: as long as I get a decent amount of time on prosthetics, five years would be enough, but end up having to have the implants taken out – even if I have to have a higher amputation, I'm simply back in a wheelchair where I started. So I weighed it up. It seemed worth it.'

'Must be amazing, feeling the wind on your stumps,' I said.

'Yeah, it's nice; it feels like being naked – like going commando. None of the sweat and rubbing of sockets. And there's no slipping. It's as solid as rock. My sockets used to move about and fell off, and I was always having to adjust them.'

'But there's this thing I can't get my head around,' I said. 'It's the fact that you permanently have a hole in you, there's a metal rod sticking through the stump – out of you. I always wondered about getting into bed at the end of the day and not being able to remove that final piece that isn't you. I think that would bother me.'

'Now it's like part of me,' Jack said. 'When I tap on it, or even brush on something, it feels just like knocking the end of my kneecap. Hold on.'

He went to get a tool.

'So it's a four-millimetre hex key, and you tighten the grub screw that bites down and attaches the prosthetic.'

He gave the hex key a half-turn and the prosthetic leg came away, and I could see the rod and the hole where it entered his skin. The naked flesh of his stump and slight redness of the skin graft sagged around the implant. *This was the future.*

'That's the fail-safe.' Jack was feeling around a small cylinder on the stem. 'It stops too much torque or force going through the implant and the bone. It breaks before your leg does – or the implant does.'

'Has it ever broken on you?'

'Once or twice. If I have a bad fall, it just breaks these little pins inside the cylinder and the prosthetic spins round, rather than my bone spinning around. It's a bit of drama. You need a toolkit to fix it, and spare parts.' Jack was holding the implant. 'They're pretty sharp. I've ripped a few bed sheets – expensive ones. So I now put a couple of tennis balls over the end.' He stroked Bruce, whose eyes had never left him.

'And no one sees it any differently?'

'It's a bit ugly,' he said. 'But then stumps aren't the prettiest, are they?'

'Nope.'

'To be honest, there's so much feedback. It goes straight into your skeleton. If I'm on gravel I can feel the vibrations going up. I was like "wow" when I first experienced it … the crunching, it's just so responsive. There's no movement between me and the prosthetic; it's directly fixed to my skeleton and much more stable …' Jack tailed off. 'The surgery is pretty brutal. I could put you in touch with the surgeons.'

'That would be great,' I said. 'So, no regrets then – you're happy with it?'

'Before I had it, I was like: fucking hell, this is drastic. But now, because of what it's done for me, it feels normal. Within two days of surgery they got me up and weight-bearing. Increased it gradually each day. By discharge from hospital, I was almost putting my whole weight through the implants – after only a couple of weeks. It was quick.'

Later, we went to lunch in a local pub. I said I'd pay, so Jack said we'd go to the more expensive one. It was great watching him walk. I'd seen him walk before, in rehab, but only between the bars in the fitting room, or for short periods in the gym. Now, nearly ten years after injury, when most people would have given up hope of any progress, he was walking confidently. I'd have to change the way I pictured Jack: no electric wheelchair. After lunch he gave me the surgeons' numbers and I drove home.

In the shower, I try bouncing my body up and swiping my arm at the lever, but only knock it half-shut and slip on the shower tray as I land and bang my shoulder. I shout in frustration. I give up the experiment – it's no fun – and use both arms to turn the shower off and swing myself out, pull a towel off the rail and drape it over my head and sit there, hunched.

I used to get angry at people who thought we were all the same. That somehow one amputee suffered the same challenges as the next. I tried to explain: *We all rehab at different speeds, like any recovery*. But even I, who was right at the heart of it, would sometimes look over in frustration at Jack, struggling to walk, and think: *Come on, just try harder; we all find it painful, why can't he do it?* No matter how much empathy we have, it's difficult not to group disability together. We talk of spectrums and severity, yet our brains simplify the world and we lose the nuance, assuming that those with the same disability will have the same challenges and experiences. Trying to shower with one arm reminded me of how differently Jack and I experience being amputees. The gulf between our disabilities is as wide as the one between me and an able-bodied person – maybe bigger.

I look at my legs. Trying to take them off with one arm had been hard enough, but I knew putting them back on would be harder: rolling a rubbery liner over the stump, getting it positioned so there were no air bubbles, so it didn't rub and the leg stayed on, could take many attempts. Jack had told me as much. Donning prosthetic liners and sockets was often frustrating to the point of despair for a triple amputee.

I understood why osseo had been life-changing for Jack: seeing how he'd attached his legs with just a turn of the hex key; how well he walked; how much healthier he looked. One of the last things he said to me as we sat together having lunch was, 'Osseo pushed me to lose weight. I felt like a bum before. I feel like more of a person now, and so much more independent – I'm more positive. More upright. I even crap better.' I'd laughed at that.

I feel over the end of my stump and try to imagine a metal rod sticking out. I imagine holding on to the implant and moving it around, and my whole stump moving with it – a direct link from outside me to inside. I can't quite do it, there's something otherworldly about the possibility. Life-changing had meant body-changing for Jack. The difference was also in the language: Jack now attached his legs, I still put my legs on.

Why does the thought of osseo still make me uncomfortable? Yes, it's extreme, but aren't the changes to my body already extreme enough – why not go another step? It's not as if I'd be ruining something that is beautiful. Familiarity has desensitised me to the ugliness of my injuries, yet I know there is something of the Elephant Man about my body. I sit, bent over my stumps, and try to see them afresh, in the same way anyone might pause to consider the strangeness of their body, the weirdness of a human foot or nose or ear. What's not there is most striking – the absence of legs – and it flicks a switch in our expectations of normal, it's a space that should be filled. Then my funny little right stump, ten years of muscle wastage making it much thinner than a thigh should be, with its deep trench of a scar running up to my buttock, so deep I can put my fingers into it and feel the tough ridges of scarring; and my left leg, which has a strange fold where the surgeons sewed up my skin – like the corner of a stitched cloth bag, leaving what the doctors called a *pig's ear*. And it does look a bit animalistic – when I bend my knee, there's the *chestbuster* that bursts through John Hurt's stomach in *Alien*. It reminds me of that.*

The first few times I studied my body, in those early months of recovery, I cried with sadness. I tend not to look

---

*I should remember the times I share a bath with my toddlers and they hug my stump and call it a dog, and even kiss it, and I am filled with a feeling that I can't describe; it is unconditional love, blind to any cultural influence of what ugly might be, and I experience a sensation of utter acceptance.

so closely now – it doesn't seem very good for my soul. I know my injuries make people look twice in the street – they see the sleek tech – but it's only when I take the legs off and peel away my liners that the full extent of what a bomb can do is revealed. Very few people have seen that. If *scars tell the story of your life*, then a bomb writes an ugly chapter. Yet I'm lucky; I can squish it all inside my prosthetics and hide it under the technology. Some people can't hide their disfigurement so easily.

I dry myself and wrap the towel around my waist and bum-shuffle out of the bathroom. I don't put my legs back on, it's nearly bedtime. I watch my partner from the doorway – she is in the kitchen, leaning against the counter on the phone. I'm reminded of those film scenes where the hero's past is revealed. They have removed their shirt, or are getting out of the shower, and the love interest sees their disfigured back for the first time and runs their fingers over the ridges of perfect make-up-artist-created diagonal scarring. Somehow the hero is now sexier and their connection deepens. I've always hated that – bombs don't do sexy.

A few years ago I was invited to the International Society for Prosthetics and Orthotics conference. It was a day of dry lectures. Each researcher, doctor or prosthetist stood up and told the audience of researchers, doctors and prosthetists their latest thoughts on socket developments. Lots of in-jokes and back-slapping. As an amputee, I seemed to be the odd one out – like being one of the ingredients at a chefs' convention. It wasn't the stare of strangers; it was the stare of experts, looking at my gait, alignment and set-up. *You know, you shouldn't pair that foot with that knee, it'll invalidate the warranty.* There were also a number of people who had knelt in front of me in the casting rooms of limb-fitting centres and wrapped plaster sheets around my stumps and pushed their hands up into my groin, looking for my ischial bone.

At the end of the day, with the room now half-empty – many attendees having already shuffled out of the rows, professional development points already accrued, mouthing 'last train' to their colleagues – the final slide quivered there, a stock image of a silver bullet above the words: *There is no silver bullet.* During the day we'd seen a series of presentations about the latest advances: pressure casting, MRI scanning, 3D printing, an hour on the benefits and risks of osseointegration, a few different lectures on NHS costs and failure rates.

The problem they were grappling with – whether you used the 'old-fashioned' craft of hand-casting with plaster-of-Paris gauze applied by a prosthetist, or the latest imaging and fabrication methods like the SocketMaster, a huge EU-funded machine that pushes multiple pressure-sensors up against the stump to take a 3D model – seemed to be twofold.

First, none of these methods could take a cast of the stump under the sort of dynamic loads and movement an amputee makes when wearing a leg in real life. Casting could only be performed static. As soon as you walked about, some allowance would have to be made for the body in motion and the socket would need to be adjusted. No technology could yet replace the dialogue between an experienced prosthetist and an amputee who could communicate what felt wrong.

Second, how do you attach a piece of machinery to the body so that it can withstand the loads required for it to be useful, but not cause pain and damage? With my sockets, there always seemed to be a compromise – to achieve a tight enough fit, I had to accept that there would be pain and the chance of some damage. The best alignment, which meant I could walk correctly, put pressure on my joints and bones – the risk was a future of osteopenia (loss of bone mass), arthritis or worse. For all the technology available to us, the prosthetic socket was still wholly suboptimal. It was only osseointegration – removing the socket altogether – that seemed to go some way towards solving the problem, yet it held the skin open for ever. In return for being able to walk again, the boundary of Jack's body had to be held open per-manently – that was the risk he had chosen to take.

Skin is the physical container of our body, and the bound-ary between ourselves and the world, protecting us from the assaults of physical injury and radiation; from the invasion of bacteria, fungus and virus; from the chemical threats of irri-tants and allergens. It insulates us, and constantly adjusts our sweat and blood supply to regulate our temperature. It stops what is inside us getting out.

Our very first understanding of the world was formed through the sensations of skin against skin: this mother–infant bond the start of a lifetime of interactions. It is essential to the way we navigate the world and is startlingly sensitive

– a particle of grit rolled between the fingers, too small to see, can feel like a boulder. Close your eyes and it will feel even bigger. By sensing pressure, heat, pain, vibrations and surface, we adjust our bodies to what is around us. It's one of the crucial ways we interact with tools – we grasp steering wheels and tap keyboards and push buttons, we manipulate and get feedback, we control our smartphones using the conductive properties of skin to change the electric field in the transparent conductor sandwiched in their screens.

And skin communicates so much of who we are and how we feel. It is one of the main physical attributes that determines how we see ourselves, and others. Sunspots, texture and wrinkles show how old we are and how hard a life we might have lived. Skin's pallor reveals how healthy, tired or dehydrated we are, and if we smoke or drink. It can prickle with fear and flush with arousal. We have evolved to be almost magically adept at reading meaning in the non-verbal language of skin's moving surface: the faintest blush or crease, signalling fear, attraction, confusion, sadness, embarrassment. *I read him like a book: thick-skinned; thin-skinned. He got under my skin …*

We alter skin. We exfoliate and moisturise it, peel it, mask it, cup it, cover it in make-up, lighten and darken it with dyes, inject it with Botox. We want it to be flawless. We mark it with tattoos and pierce it. It carries social, cultural and religious meaning. Some of the greatest human suffering in history is because of its colour.

And if you want to break someone's mind, you go to work on the surface of the body, scarring, burning and disfiguring the skin to find a way to the soul.

Every now and again – once a month, say – there is a day when it feels like I am inside a cruelly devised torture machine. These are the days when the liner plucks at my pubes until the skin is pulled red and spotted with whiteheads; or the socket edge rubs my groin to a raw, open welt; or my bone jars against the socket wall; or some pressure sore can't heal inside the moist sweat of the silicone and is hot with inflammation – with blisters, sweat rash and ingrowing hairs. Or it's just the sensation that I have spent the day walking through treacle, condemned to some purgatorial punishment of a Greek myth.

It is damage to my skin that is so often the reminder of my union with technology. It becomes the site of pain and risk – the messy edge where unforgiving technology meets skin and flesh. We all have a sense of this, and while rubbing shoes, itching contact lenses and uncomfortable earphones are an irritation, the musculoskeletal dysfunction or chronic skin breakdown and infection caused by the prosthetic interface are more serious.

The Deal

A few days after seeing Jack, I'm given the coffee I order for free. 'It's on the house,' the woman says. I wonder if my prosthetics mean I am given this perk more often than others. I sit at the back of the café and open my laptop. There is someone approaching – it's the barista, and I think I've left something at the till. She limps.

'Sorry. Do you mind?' She has an Eastern European accent. 'I saw your legs.'

'Oh yes?'

'I wondered if you can help me. I have a bad ankle. And I might have to have it amputated. I'm just wondering what it's like for you?'

I'm asked this surprisingly often. Someone has an injury (or their cousin does) to their knee or ankle that won't get better, cartilage is wasting away or a bone won't heal, and they need surgery. A doctor has mentioned the possibility of amputation. Sometimes their mobility is so reduced that they've been told prosthetics might give them a second chance. Most of all, they are in pain. But it's hard to know where to start.

'Sorry to hear that,' I say. 'Thanks for the coffee.'

'Your legs – they look amazing, and you're so fast on them.'

'It's okay. It takes some getting used to.'

'How long?'

'I could do it without a stick after about four months, but for it to feel normal, probably a year or two. I was lucky, I was given the best prosthetics and had all the rehab I could need. I was injured in the military.'

'How does it feel?'

Describe the indescribable: what does it feel like to wear a prosthetic? Time spent as an amputee seems to numb the inescapable discomfort of it. It's a bit like wearing a wetsuit material over your legs without the relief of getting wet, so it's hot and chafes; and like wearing too-tight shoes soled with lead, which seem to tighten during the day while at the same time becoming loose, so that they slip and bump your toes and heel. I simply say, 'It can be uncomfortable.'

'So it hurts, yes?'

'Yes, sometimes a lot – all I want to do is take them off. Some days it doesn't hurt at all. You get used to it.'

'You think I'd be okay?'

I couldn't answer this, but ended up replying with some non-committal, upbeat pep-talk. In truth, no one could give her an answer. There was the risk of surgery, the lottery of phantom and nerve pain, the whole long, incremental process of socket-fitting and refitting, the years of becoming accustomed to the pain. The seemingly random flare-ups and complications that could turn you from independent and pain-free to suffering and immobile. It's why doctors and patients tend to try everything to save a limb, before taking it off. But then so often, when people are given renewed freedom after amputation, they say the years of trying to save a damaged limb were wasted time.

I thank her as I leave. Yes, it's okay today. In fact the prosthetics feel comforting, cocooning my stumps, and I'm walking on a cushion of air – it's almost fun. If it had been a day of pain, maybe I wouldn't have been so upbeat; maybe I would have told her to listen to the advice and not rush into anything. It's no picnic, I could have said, but that might have been lost in translation. I know she's still watching me walk down the street, trying to imagine.

In her book *A Heavy Reckoning*, Dr Emily Mayhew charts the medical history of the Afghanistan campaign. She describes how advances in medical science, techniques and technology managed to save soldiers who shouldn't have survived, and she writes about 'The Deal' – that, like the myths and legends found throughout human culture and history, making a deal to cheat death has consequences. When you bring a body back from the brink there is always a price. For soldiers like me, she says, the price will be the 'conditions associated with old age: hypertension, diabetes, coronary artery disease and chronic kidney disease (just for starters – there are probably others)'. I will age faster, and my life has been shortened.

Everything that happened in the hours and days after injury was a negotiation: life now, in return for less life and poorer-quality life later.

I hated reading this. I was determined to prove her wrong – I would live a long and healthy life.

But I knew there was truth in what she wrote. Being saved did have costs. Even for amputees with the most successful rehabilitation, the relationship between body and tech is one that needs constant attention. And it's not only us amputees; repairing the human body when it fails almost always has costs. We're told about the medical side-effects – they're on the back of the packet, or are read to us as we sit in waiting rooms – and we sign for ourselves, or on behalf of a loved one, that we have understood and consent. The extent to which we pay attention to those risks depends on how great our need is to be fixed, or to be pain-free, or to have a few more years alive. So often, when we fix the body with drugs and surgery and prosthetics, there is a trade-off and we enter into a relationship that has a price.

Watching a Parkinson's sufferer's uncontrollable tremor reduce almost to nothing as a doctor 'turns a dial' on a computer is astonishing. You can watch the clips on YouTube: the patient sits in a chair beside the doctor's desk, shaking violently. The doctor is about to turn on a device implanted deep in her brain. We're shown footage of her a few weeks earlier. The patient in the operating theatre, awake, an infection barrier of skirting around her forehead, and a group of surgeons standing behind, having drilled a 'coin-sized' hole in her skull. They push an electrode down towards the centre of her brain. Another medic talks to her, watching for any tripping of her responses that will indicate they are about to damage crucial parts of her brain.

Deep brain stimulation (DBS) is like a pacemaker for the brain. It is used to treat diseases such as Parkinson's, essential tremors, dystonia, epilepsy and obsessive-compulsive disorder – and, if placed in other parts of the brain, even depression. The device consists of an impulse generator (a battery pack) implanted near the collarbone, a wire running up the neck to the electrode, and the electrode itself is placed in the centre of the brain, near the thalamus. It's all enclosed under the skin. The device produces high-frequency electrical stimulation that overrides the abnormal brain signals that are causing the involuntary movements. (We have theories on how DBS works, but it's an example of a medical technology we don't fully understand; we just know that it does.)

Back in the doctor's room, the patient is about to have the device turned on for the first time. Before the big moment she is asked to do a series of tests: touch her nose, touch the doctor's finger, bring a cup to her mouth. She can't do it. The shaking is uncontrollable. Then the doctor turns on the device. Deep in the brain the electrode sends electrical pulses into the neural tissue. The doctor starts to change the frequency until the shaking almost completely stills. The patient brings a cup to her mouth and drinks, writes her name, buttons her shirt. She hasn't been able to do any of this for years and cries. Given how invasive the operation is, the outcomes are overwhelmingly good, side-effects are rare and, while DBS is not a cure and symptoms often slowly return, there is no question that this surgery can be life-changing.

During the last fifty years developments in materials, batteries and electronics have resulted in a huge range of indwelling medical devices designed to benefit diseases relating to every organ. Joint replacements, meshes to support organ tissues or defects in the abdominal wall, intraocular lenses for cataracts, cochlear implants, dental implants, pacemakers, valves and stents and artificial hearts, spinal-cord implants, brain implants – they are all becoming increasingly common, and a surprisingly large proportion of us (twenty-five million in the US alone) now rely on these invisible, artificial body augmentations.

But the body is not a bystander in all this – there's a trade-off at the biological level. You only have to experience the wincing pain of sand in the eye, the redness around a splinter before it pusses out, the convulsing throat when you have swallowed something you shouldn't have, to know how violently the body can respond to a foreign object. As I healed after injury, the shrapnel that flecked my arms and legs activated the foreign body reaction (FBR). My tissues first tried to ingest this debris (using phagocyte cells) in an acute inflammation response; and when they were too big or too metal to be ingested or ejected, fibroblast cells walled them off with scar tissue, creating a granuloma or cyst, and isolating them from the rest of the body. This shrapnel may stay locked under my skin until I die; it may migrate to the surface, to be picked out one day; or it may get knocked in the future, causing an infection that will need antibiotic treatment.

And indwelling devices are no different. Yes, they are made of carefully chosen materials that help minimise the intensity of the FBR, and there are rigorous sterilisation measures during surgical insertion; but, as with any foreign body, there is a profound effect on the tissue implants are placed into, and they can trigger the host's inflammatory response. The devices are treated as alien and walled off. And in the gap between wetware and hardware – however small, however clean on insertion – there is the risk of infection.

Bacteria live in two ways: either planktonic or as encapsulated communities called 'biofilms'. In planktonic form, they float around in solutions as isolated individuals. Like this, they are vulnerable to antibiotics and the immune response of the host. But bacteria are hardier when living in biofilms, mini-ecosystems that have been described as 'cities for microbes' – they benefit from strength in numbers, share genetic material and, critically, build themselves into three-dimensional structures that make them resistant to the host's immune response and hard to treat with antibiotics. (The plaque on your teeth is an example of a biofilm.) And once the 'city' is no longer beneficial to the 'success' of the community, chunks of bacteria may break away, spreading infection around the body that can lead to septic shock and death. The gap between the body and a foreign object is the perfect environment for these biofilms to grow.

It's a mark of what is at stake that implanted medical devices are now the most common cause of healthcare-associated infection (at between 50 and 70 per cent); it's also an indication of just how many of us have something implanted. These infections need treatment with antibiotics, surgical revisions or removal of the implant. And, depending on the device, there is a risk of death. If your urinary catheter gets infected, which 100 per cent do at some point, the risk is less than 5 per cent; if it's your mechanical heart valve, the risk is more than 25 per cent. And as the threat of anti-microbial resistance increases, we have fewer ways of dealing with it. It's a problem of evolution: there are many more human hosts for bacteria to infect, and they become more resistant as they mutate to overcome our antibodies and antibiotics.

We tend not to get shown this side of medicine. We want to trumpet our successes from the rooftops and not dwell on what we cannot control – and controlling the inflammatory reaction of the FBR, and the infections that come with it, is still a struggle. It's a frustrating side-effect, outweighed by the brilliance of the clinical solution implantable tech brings – we can literally save life by replacing and augmenting the body. But as we create more sophisticated devices, which interact with the body in new and subtle ways, the FBR becomes a greater problem.

We marvel at the bionic eye: a device implanted into the brain that can excite the neurons in such a way as to give a visually impaired person some sight. The device is shiny and futuristic – there is light where once there was darkness – yet the press articles don't mention that the body begins to attack this state-of-the-art invention. Current bionic eyes will have, at best, an array of 100 electrodes placed carefully on the cortex. This results in the patient seeing a flickering ten-by-ten matrix of dots; the perfect patient may be able to make out shapes and some difference in edge, but the resolution is far below anything that might be counted as restored vision – it's more of an aid to navigating the world.

In these types of devices (which interface with neural tissue), we historically used metal electrodes – they are non-toxic and can carry the charge to excite the neurons. But for the patient to see a spot of light (or a sound perception, in a cochlear implant), the charge has to be fired at a certain strength. The greater the current used, the more damage you do to the brain tissues that you are trying to activate – tissues don't like being zapped (at the molecular level there are all sorts of things we don't understand about the way the charge and spin of particles is critical to the proper functioning of cells – and adding a dose of 'prosthetic' current upsets the balance). While the patient may see the spots of light clearly when the device is first turned on (think crying in the doctor's chair), the body is already walling off the device, insulating it from the tissues that need to be activated. So more charge is required to push through the scarring and get a spot of vision, and the device is 'turned up', which further damages the tissue, causes greater scarring. Over time, neurons die and the device becomes less effective.

But visually impaired people want these high-tech devices to show them more than a few flickering dots, which might just about be an aid for navigation. (There are other, smarter ways to find their way around: the white stick, the guide dog.) If you're going to have electrodes implanted in your brain, you're going to want it to be worth the risks. And if we want to make the kinds of devices that connect with the nervous system to replace sight loss with more than simply the most rudimentary spots of vision, then the interface between hardware and wetware will have to carry more information more precisely.

Because the metals of existing electrodes are stiff and inorganic – prime candidates to be treated as a foreign body by human flesh – researchers are developing new, clever materials that can sit more symbiotically within the brain. They mimic the tissues, reducing the FBR while still being able to carry the current needed to create a sensory therapeutic benefit. One solution being tried uses hybrid materials: electrodes coated in hydrogels seeded with stem cells that can grow synaptic connections between the device and the nervous system – essentially a living electrode, with a softer coating and less material mismatch between technology and living tissue. If the electrode is made of the same stuff (or nearly the same stuff) as us, there is less chance of an FBR and a better chance of more natural communication between the device and the body.

I've had a number of infections since becoming injured. I will have been in good health for years and then I'll feel a bit odd and an area of one of my stumps, or an old slither of shrapnel in my arm, will flare up. My flesh will become hot and painful. Antibiotics will sort me out, but the pain is significant. It's also mixed with the anxiety that maybe this time the antibiotics won't work and they will have to chop off a little more of my leg. And even when I'm in good health, there is constantly some level of background pain. It seems to be the trade-off of hybrid living.

Ten years as an amputee has made me (think of myself as) a bit of an expert on pain. If I'm asked what wearing prosthetics is like, the first thing I talk about – perhaps with a little too much of the martyr – is pain. In the early days, when the pain was new and worryingly alien, I swallowed a palette of colourful pills and they did help, but *the deal* was that everything was experienced through a fog of nausea. So I made a conscious effort to change my relationship to pain, to try to live with it and come off the meds. (Too quickly, as it turned out, leading to an appreciation of what 'cold turkey' is like; and as I bent over a sick bowl, looking at the anti-nausea tablet I'd swallowed a few moments earlier floating in a puddle of bile, I begged the nurses to give me back my drugs.) In the end it took much longer to 'go clean' and I could only manage a gradual reduction. I don't take pain-meds regularly now. I've learnt first-hand that the terms of the deal we have to enter into, when we take these drugs, isn't worth it.*

---

*The cost of painkillers is well documented at the population level – in many countries there has been an opioid crisis of over-prescription and misuse. In the US the number of deaths related to opioids increased by 90 per cent between 2013 and 2017, rising from about 25,000 to more than 47,000. There were 1.7 million people addicted to prescription opioids, costing the country $78.5 billion each year. It is such a costly problem that new prescription guidelines have been introduced for doctors, and more funding has been put behind alternative pain-treatment research.

We have alternative pain therapies: peripheral nerve-stimulation therapies that get right into the wiring of the nervous system. These indwelling devices work in a similar way to DBS – electrodes implanted next to the spine (or a nerve in the roof of the mouth, to target cluster headaches). They modify the pain signals so that the patient doesn't feel them, or feels them as different, less painful sensations (sometimes called paraesthesia), like a tingle or pins and needles. Then there are softer solutions, some of which feel like throwbacks to witchcraft: acupuncture, transcranial magnetic stimulation, alternatives to pharmaceuticals such as cannabidiol, and virtual reality. Whatever the therapy we use, pain is still one of the hardest conditions to manage, and one of the most under-treated.

For everyday purposes, we all know what pain is: we burn our hand, it hurts, we need to make it stop hurting and we yank our hand out of the hot water. Put a little more scientifically: noxious (thermal, chemical or mechanical) stimuli to the sensory nerve cells in our tissues, called nociceptors, cause a signal to travel along a series of nerve fibres via the spinal cord to the brain, triggering actions that reduce and regulate the pain (the hand pulling away, and the release of endorphins and enkephalins – the body's natural painkillers). This is nociception: the nervous system's way of protecting us.

But we also all have a sense that nociception alone isn't pain – pain is also the transformation of those bodily mechanisms into an emotional experience. We feel sad, angry and frustrated that we've burnt our hand. And while nociception is observable (we see the flinching hand, and can use

69

instruments to measure signals firing in our nervous system), the emotion of pain is not. It is completely subjective. It's no wonder those pain metaphors – *it's like a sharp, stabbing, crushing pain* – and 'Please rate your pain between one and ten' are the only ways we have to tell the doctor what it feels like.

Knowing how pain works seems, to me, to be far less important than how much it hurts, or how long it will last. Pain caused by a specific injury or disease, which serves a useful protective purpose and then disappears as the body heals, is called acute pain. Pain that outlasts the normal time of healing, often defined as lasting more than three months, we call chronic pain.* This is when pain becomes a disease in itself. Why pain becomes chronic isn't fully understood. We know that after an injury the pathways of neurons in the spinal column develop a heightened state of sensitivity. This is protective – part of the body's way of reminding us that the hand needs looking after while it heals. But sometimes the neural pathways become hypersensitive, and remain so long after the injury heals. One theory is that a barrage of pain signals from an injury can amplify the message on its journey through the spine to the brain. This is sometimes

---

*Because it so often gets confused, it feels important to note that the word 'chronic' isn't used to describe the severity of pain, but simply that it persists over time. A long-lasting pain that is mild would still be chronic. Even so, chronic is shitty because there is no end in sight, and that's painful in itself, so we get confused. (A new term being coined is 'enduring pain'. It better describes the fact that the pain persists and has to be endured.)

called central sensitisation or 'wind-up'. The pain is like an orchestra playing a tune: if the tune keeps playing, the nerve pathways get better and better at *playing* and continue to play the one tune they have learnt, long after it's useful for them to do so. This, as far as we know, is what happens in patients with conditions such as fibromyalgia, irritable bowel syndrome and other forms of neuropathic pain.

And the longer we have pain, the more emotional it gets. It holds our attention, isolates us socially and stops us sleeping. We ruminate and catastrophise, and the anxiety surges. It makes us miserable – especially if it's no longer linked to some injury that we can see, or the doctor can explain. Pain can become the disability. For many disabled people this is their reality, one they must come to terms with. The disabled feel more severe pain, and more often, than the non-disabled. This can be a direct result of an atypical body or a secondary condition: the pain of inactivity leading to obesity; bladder and bowel problems; musculoskeletal abnormalities; and strange neuropathic pain. Or the pain of interfacing with technology – the side-effects of drugs, the pain of routine medical procedures, the pressure sores of wheelchair use or rubbing prosthetics. To be a hybrid human is perhaps, for now, to experience pain.

It's worth saying that being an amputee does bring a special pain that most people will never experience. Phantom pain is pain that feels like it is coming from a body part that is no longer there. Sometimes it is not pain at all, but phantom sensations. It's often triggered by certain events: cold weather, for instance, or emotional triggers or (one of the most unpleasant, in my experience) forgetting you've lost the limb and trying to use it – imagine impulsively trying to kick a ball that bounces towards you, but the leg isn't there: the ultimate air kick.

I once shared a hospital bay with an amputee who complained that his missing leg was frozen in an uncomfortable position, so that as he lay there, it felt as if his knee was bent down through the bed. This was strange and painful, and he couldn't sleep on his back. And I met a single-below in rehab who was in a burning hell so bad that he had surgery to revise the nerves in his stump.* Another described his pain as like being invaded by crawling insects. And a historical example: after Nelson lost his right arm in the Battle of Santa Cruz de Tenerife (1797) he said it felt as if fingers were digging into his palm. For him this was proof of the soul – if his arm could disappear and still feel, why not a whole body?

What does a hot-itchy-stabbing-cold-pulsing-electric pain mean? My unique experience won't be very useful, but I'll try to describe it anyway. First, anatomically, when I am wearing my prosthetics, my feet descend to inhabit my

---

*To stop the phantom pain, he had a revision of his stump. The surgeons opened him up and found the neuromas – the painful clumps of nerve that often develop when a severed nerve tries to regrow – and, using targeted muscle reinnervation (TMR), grafted two severed nerves together to create a new connection that prevented fresh neuromas growing and stopped the pain.

carbon-fibre feet and tingle; but when I am in bed, they tele-scope up, so they are at the end of my stumps and become small and prickle. Some people talk of being able to move their phantom limbs, but I can't; my legs and toes are fixed solid, and no amount of trying to move them does anything. Trying to can even trigger the pain, which builds the more I think about it – it's best not to. And, perhaps most weirdly, as I write this, my limbs are becoming painful, as if I am engag-ing the pain pathways ...

I can feel the bridge of my right foot now; it is there and the physicality of the skin is a sensation that makes me see the wrinkled whiteness of that skin in my memory. And my toes too are hurting, as if scuffed on a sandy beach and the salt is stinging them. And when they really hurt, once a month or so, it is in the dead of night, and hot shocks pulse and blip like bubbles through lips – it is an electric fence I can't let go of; and sometimes it is a very thin knife stabbed into me and left in the limb to become a dull ache ... But it's always dif-ferent and surprising. It can make me laugh, it can make me cry out involuntarily, it can force me from bed to put my legs back on for the relief of the pressure of the sockets.

Then there is the un-itchable itch, down in my calf, which comes every few months, and there is no solution but to endure and enter – as Dante described it – the eighth circle of hell, reserved for the very worst sinners, who were punished to suffer 'the burning rage / of fierce itching that nothing could relieve'. And sometimes, if I step in a puddle while out walking with the kids, my feet get icy cold and I can't get them warm, even hours later, and it is ice-block painful.

Most strange is the huge wave of pain that fizzes from the soles of my feet when I orgasm – apparently due to the way our nerves are wired through the spine – a very intense mix of pleasure and pain. Whenever I've talked of any of this to doctors or scientists, they have nodded in understanding, but as soon as I ask why I have pain so long after injury or how it works, they start to shake their heads – they're not sure. Pain is poorly understood. Pain is strange. Pain is part of the deal.

For all the pain and the risks of side-effects and the FBR and device failure, I believe there is something more hopeful that comes with having to live with the deal. I can only speak of my own experience, but despite the many costs that dependence on med-tech brings – the pain that can make me irritable and short with my children; the anxiety that has ruined holidays; the frustration that I can't run for the bus, or play football, or dance all night – I also have a new experience of the world I wouldn't otherwise have felt, and I have a feeling my life has more meaning than it once did. We know med-tech can change a personality: drugs can do it, and some people who have DBS report a change. And I too feel like I am changed for the better.

So often, all people can see is the deal. They look at a disabled person and assume they are experiencing only their disability, when actually it forms very little of their experience – they are more than just their disability. This won't be true for all people. I know I've been lucky that the disability I acquired allows me hope, and to feel that, despite the costs, I wouldn't want to change it. It's hard to describe why this is.

Years back, I had an exchange with a woman. I'd been talking to a group of people about my experiences of injury. It went something like this:

She asked me, 'How did you deal with your PTSD?'

I replied, 'I've never had PTSD.'

'Of course you did.'

This pissed me off. 'How can you tell?'

'Well, I was a psychotherapist – and by the way you talk about it all. Look at you.' She waved a hand towards my prosthetics.

As much as I tried to tell her I hadn't had Post-Traumatic Stress Disorder, that I was lucky and even accepted that PTSD might one day rear its head – there is often a fifteen-year post-trauma spike in cases, which I might fall prey to – she simply smiled and looked at me in sympathy. I felt powerless in the face of her conviction. I told her how everything that had happened to me actually felt overwhelmingly positive. I tried out some of the science I knew: that we all experience Post-Traumatic Stress after a trauma, it is part of the natural response; it was the D – the Disorder bit, which was a malfunction of the brain – that I hadn't had. She nodded, and I thought she was going pat my thigh and say, *There, there*. What was so irritating was that she saw me as a victim, but getting angry with her seemed to be playing to her point, so I moved the conversation on.

The narrative around PTSD often means we don't realise that something else might be possible. Around half of us will experience a significant traumatic experience in our lives, and fewer than 10 per cent of those who do will suffer

PTSD. Post-Traumatic Growth (PTG) is the positive change someone can experience in response to a life crisis and, although it is by no means guaranteed, it is actually more common than PTSD. The growth can be felt in a number of ways: a new appreciation of life, a deepening of personal relationships, becoming emotionally stronger, changing one's priorities and having a fuller spiritual life. When we hear this – that good comes out of suffering – it seems obvious; it's an idea as old as civilisation.

Some people are better equipped to feel growth: support from strong social networks, from friends and family, and being healthy and financially comfortable helps – as do personal attributes such as resilience, optimism and being easy-going. It's important to say that traumatic events are not a good thing in themselves; they often result in a feeling of life's futility. It's complex: PTG and PTSD can often coexist after trauma. You can feel lower self-esteem, trust others less, let yourself go, feel more vulnerable and still have a new appreciation of life.

People often feel Post-Traumatic Stress when diagnosed with cancer, but once treatment starts, they feel growth, life takes on new meaning and they have a closer connection to those around them. This is how I feel. And I've always had a strong feeling that the technologies that fixed me were vital for my own growth – for the euphoria I feel, to still be here. Assistive technologies have given me a way out.

Pass the Bone Saw, Please

On a bright late-summer day I travel to Birmingham to visit the surgeons who performed Jack's osseointegration operation. On the train I read a few journal papers about the procedure. I know I can't have the operation; even if I could stomach the idea of a metal rod holding my skin open for a lifetime, I don't qualify – I wouldn't pass the stringent ethical protocol. I put on my legs at seven in the morning and they stay on until I go to bed. For an amputee, that's good going, and the risks of osseo aren't worth it. However, I can't help but be fascinated. I need to understand it better – and in years to come, when my body has deteriorated with years of wear and tear, perhaps osseointegration will be an option for me. Just as Jack said, the cons, the otherworldliness of it, will be outweighed by what it can do for me.

The first article is a review of research into osseointegration. Thinking I'd be reading about amputees, I'm surprised that the focus is dentistry (and that the technology is more than forty years old). I hadn't realised the technique was most commonly used to fix dental implants in place. A hole is drilled into the jaw and a titanium 'root' is screwed in. The new prosthetic tooth is then attached to the titanium screw, which integrates with the bone. I pick up my phone and search for pictures. I scroll through the thumbnails of open, bloody mouths mid-surgery, or the tops of screws sticking from gums and topped with shiny teeth (there are a few brown necrotic-looking failures and infections in the mix). The paper goes on to outline other uses – bone-anchored hearing aids, attaching maxillofacial prosthetics, such as ears and noses, and finger and thumb prosthetics – but it doesn't mention lower-limb amputees. There's a sentence or two on the how the procedure was invented.

By accident, it turns out. In 1952 Per-Ingvar Brånemark, a Swedish orthopaedic surgeon, was experimenting on rabbit legs, trying to observe the blood flow in living bone marrow (only dead bone marrow had ever been observed). To do this, he drilled titanium implants housing microscopic lenses into the rabbit's legs, through which he could make the observations of the marrow over a few months. At the end of the experiment he wrote a long paper outlining his findings and then tried to remove the titanium implants. He couldn't – they'd fused with the bone. He'd stumbled on the discovery that titanium could structurally integrate into living bone and didn't irritate soft tissues, either. After successfully experimenting with titanium dental implants in dogs, he suggested the procedure would be a long-lasting solution for anchoring prosthetic teeth in people.

Because he wasn't a dentist, the scientific and medical community didn't accept Brånemark's discovery. They also didn't believe that titanium could integrate with bone without the formation of fibrous tissue other metals caused, which so often led to complications and infection. This wasn't helped by the fact that Brånemark couldn't actually explain why titanium performed differently. He was shunned and abused at conferences and seminars, his university research funding was withdrawn and there were calls for his expulsion. It was only by setting up his own private dental practice, where he started performing dental osseointegration in people, that word spread of his success and the procedure was finally adopted. He is now seen as the father of modern dental implantology.

The next article I'd printed off focuses on the 100 or so Swedish amputees who've had osseointegration during an eight-year period – one of the authors is Rickard Brånemark, who'd followed his father into medicine and pioneered osseointegration in lower limbs. The paper is a bit dry, discussing rehabilitation protocol and standardisation, but there are a couple of references at the bottom that catch my eye. What becomes apparent, as I flick through Google Scholar on my phone reading the abstracts of other academic papers, is that there are a few different ways of implanting a metal rod into a leg bone. And there seems to be a bit of competition. The front-runners are Brånemark's Swedish screw-fit stem and an Australian press-fit stem pioneered by a doctor called Munjed Al Muderis. I find a *Sydney Morning Herald* article about him.

Al Muderis grew up in Iraq, the only son of a privileged aristocratic family. But in 1999 he had to flee the country.

He was working as a young surgeon at Baghdad's Saddam Hussein Medical Centre on the morning that three bus-loads of deserters were brought in by military police and Ba'ath Party officials. The surgical team was ordered to cancel the elective list for the day and instead amputate a part of each man's ears as punishment. The head of surgery refused, citing the Hippocratic Oath, and was taken outside and shot. Al Muderis managed to slip away and hide in the women's toilet until the end of the day. He didn't go home – he couldn't, they would find him – and instead he started a journey that led through Jordan, Malaysia and Jakarta, then to Australia by fishing boat with 165 other refugees. On arrival he was held at the Curtin Immigration Detention Centre, as number 982. Ten months later he was finally granted asylum. Al Muderis went on to become a pioneering orthopaedic surgeon.

I lower the phone and look out of the train window, black verticals ticking past in the green blur of trees.

The train sways as it enters Birmingham New Street. I make a note across the top of one of the papers to 'read again' and pack them all away. I head down the carriage, holding the chair headrests to steady myself against the train jolting to a stop. I drop down onto the flat stone platform – the perfect surface for walking. I have my headphones in and I walk to the up-beat, as if the judges are watching. I'm showing off. It's another good day on my pins.

This journey feels like a pilgrimage. It's the first time I have returned to Birmingham since 2009, when I was brought back from the brink in this city, but I hardly know it. I catch an Uber outside the station, settle in for the ride and plan to take it all in – the red-brick buildings and dark pedestrian railings. The city where I was saved. But the driver is looking over his shoulder and glancing down at my legs. I know what's coming. What is it with taxi drivers? They always ask. Maybe it's because this is his vehicle, which he's embodied, and I'm fair game when I'm inside his space.

'Mr Harry?' he says.

'Yup?' I say.

'What have you got there?' He is motioning with his head.

'No legs,' I say.

He's called Baboor. And this time I can't shrug him off. He's part of my pilgrimage. He's from Afghanistan. He tells me his brother was an officer. He was blown up in the war between the Soviets and the Mujahideen – it's not clear which side he was fighting for, but I think it was the National Army. I keep missing what Baboor says; his Afghan-Birmingham accent and the rattle of his Prius make him hard to understand.

But I'm no longer looking out of the window at the city. I'm leaning forward from the back seat into Baboor's space. His brother was injured at the same age I was; it was a night operation, *bombs in the roadside. Both legs gone.* He wasn't discharged – he served on for years in an admin role. Now he is fifty-five and lives in Canada and works in a hotel. *Easy money, easy life and good prosthetics too.* They haven't seen each other for many years, but speak on Skype. It's strange that I've been to Afghanistan more recently than Baboor. Inevitably we move on to geopolitics, and I'm relieved when we turn off the roundabout and up to the drop-off area of the Queen Elizabeth Hospital.

I meet Jon Kendrew by Costa. He's wearing the light blue of an RAF uniform, with the epaulettes of a Group Captain. Uniforms aren't unusual here – the hospital is home to the Royal Centre for Defence Medicine. He shows me up to a consulting room and we talk for an hour. We're fairly sure we haven't met, and that he's never 'worked' on me. But it's a small world and there are a few near-misses – he arrived at the field hospital in Afghanistan shortly after I'd passed through.

Jon joined the Air Force as a medic with the ambition of becoming a fast-jet pilot, but his training as an orthopaedic surgeon led to tours of Afghanistan. He describes it as the Olympics of his career, an onslaught of daily trauma surgery – so many *very poorly* young soldiers were brought in off the battlefield. In the ten years that have passed he's followed them through their recoveries and has seen some struggling to get better. It was one of the reasons he looked again at osseointegration.

He never imagined he'd be an osseo surgeon – 'I'd always thought the idea of an open wound was crazy,' he says. 'But we were driven by the patients. We knew that if someone didn't offer it in the UK, they would do it anyway privately or overseas, without oversight or the backup of the NHS – the UK had to have the balls to say we were going to try it. The challenge was making the procedure safe for patients with a history of infection and horrendous injury profiles. Patients who shouldn't be alive.'

Jon explains how they needed to pick the best implant system – the one that would cause the fewest complications for blast victims. He mentions the Brånemark Swedish screw-fit implant I'd read about, and a German system. In the end, they picked Al Muderis's Australian press-fit stem, which integrated by pressing the implant into the bone through being loaded with the weight of the patient. This meant the patient was mobile sooner and needed fewer surgical steps, which reduced the risks of anaesthesia and infection from surgery. 'One reason press-fit was chosen was its similarity to a total hip-replacement prostheses,' Jon tells me. 'Although osseo seems drastic, in some ways it's no different to a hip and knee replacement.' (I'm reminded of an article in the *Lancet* titled 'The operation of the century: total hip replacement', which charts 100 years of development, the ever-increasing implant survival rates, and patients' growing expectations for high-activity lives.)

Jon travelled to Australia to do a mini-fellowship with Dr Al Muderis. 'He's done a huge amount for osseointegration and wants to make the procedure available as widely as possible,' Jon says. 'He's provided clinics in Cambodia for mine victims who are still working in the fields – he's been named "Australian of the year".' Within a few months of returning from Australia, Jon and the team had performed the first operations here in the Queen Elizabeth Hospital.

He says the surgery is brutal. 'I still find it stressful – these are the first blast patients to have the surgery. There's the risk of fat embolism.'

I have to ask what this is.

'When the patient is on the table, you have to hammer the metal stems into the femur.' Which, by the look of Jon's hand movements, is no different from hammering during DIY. 'You can over-pressurise the system and shoot bits of emboli – fat, bacteria, blood clots or debris – into the blood-stream that lodge in place and can block blood vessels.' He tells me how fragile people who have been blown up are, even years after the explosion. 'It's very high-risk, so we're really strict on who we pick.'

Jon is also aware that people around the country are watch-ing to see how the trial goes. 'That's stressful too. I know many are surprised the surgery is being offered, surprised we chose the Australian system over the others, and that we didn't wait for the US to adopt the technology first.' And he acknowl-edges that no one knows how it's going to pan out, or what having a bit of metal sticking out of a leg is going to do in five or ten years. He smiles. 'We haven't had to take out any stems yet, no one's died, no one's had sepsis, which we've seen in people who'd paid to have it privately or gone overseas. It's a work-in-progress – less than five thousand people in the world have had osseointegration in a lower limb. But when you see people who haven't walked for years standing upright by themselves only three days after the surgery, it seems worth it.'

As we head for the stairs we talk about Jack and how well he's doing, now he's had osseointegration. 'Treating patients like Jack feels like a closing of the loop,' Jon says. 'Seeing them finally up and walking, after so long, has been good for my mental health and resilience. I feel privileged to have seen that full cycle.'

I say goodbye to Jon and return to the hospital foyer. I'm left with an image. It is of Jon about halfway through our conversation. He's sitting on the other side of the table and no longer looks at me; he's lost in memory and says, 'One of the guys, I remember, I saw him in clinic back here, I realised the last time I'd seen him was in Afghanistan and we had his chest open. He'd arrested coming off the back of a MERT. As the orthopaedic surgeon, I was helping.' It's at this moment that Jon lifts his hands up in front of him and starts massaging the air. 'I'd never done it before. I'd trained on a cadaver. But this was a British soldier, and we were doing internal cardiac massage. I was holding his heart in my hands.' Jon's hands drop. 'And he survived.' He looks at me again and laughs. 'Then you finish the tour and a few years later you see him here, and you're like: Oh my God, you're alive?'

The second surgeon I meet is Demetrius Evriviades. I knew he'd been one of the surgeons when I was in hospital, but I didn't know how involved in my recovery he'd been. It's only when he asks after my parents, and describes what it was like updating them on my progress when I was in intensive care, that I realise how well he 'knows' me. Now he's telling me a story. About seven days after I was flown back I was in surgery, having a debridement (a routine cleaning and revising of my wounds).

'I remember it very vividly,' Demetrius says. 'We opened up your leg to start cleaning it out and I saw this fungus all along the blood vessels of your leg. It scared the living shit out of me – it looked like something out of *Doctor Who*, or a horror movie. Blast had driven fungal spores into your leg. They'd taken ten days to germinate. It was really scary – sort of alien. You were very close to death.'

It's strange sitting across the table from the man who amputated one of my legs. We're in the café, drinking coffee. He tells me the decision was easy. 'There are three indications to take off a leg or an arm: it's dead, dangerous or a damn nuisance. Yours had ticked all three boxes.'

It's the first time I've been told how close I was to dying, here in the UK.

Demetrius has retired from the military since he'd been one of my surgeons. He'd spent twenty-two years in the Royal Air Force, deploying on multiple operational tours and leaving as a Wing Commander. Like Jon, he served in Afghanistan. Having worked together during the war, it was natural for them to collaborate on osseointegration. The procedure spans both their disciplines – half-orthopaedic, half-plastics. Unlike Jon, Demetrius had few reservations about the operation. As a plastic surgeon, he'd been perform- ing osseointegration in the head and neck for ten years. He tells me how he removes a damaged ear or nose, puts little osseointegrated implants into the skull, then 'knocks on' a silicone prosthetic that's been made in the lab. For him, the benefits were clear.

He opens his laptop and shows me a series of photos of implants protruding out of stumps. 'Bugs love it in there,' he says, pointing to fatty skin over the implant and the redness of infection. 'It's a nice, moist anaerobic environment for them to grow. There's wound juice. Bugs love that. You get local infection, then the body reacts by creating granulation – granulation is the body's Polyfilla – and then more infection. Around half of patients will develop superficial infection on the stoma, which makes it painful and can lead to nastier, deeper infection.' One picture sums up the very real risks of the procedure – red and putrid dying flesh around the shiny grey metal rod jutting out at the centre.

He clicks to the next image. It's another stoma, but this time the fat is pared back, with a wide area of mesh-like skin around the implant. 'What you want is a nice, calm, clean, dry environment – this is one of mine.' It does look cleaner. Demetrius had wondered why osseointegration in the skull tended not to get infected. 'In the head there's very little fat and there's no movement of the soft tissues around the implant,' he says. 'I realised that we needed to create that in the stump.' He points to the picture where the skin abuts the implant. 'We used synthetic biological skin substitute to grow over the bone. It's called Integra; it's made of a mix of cow and shark cartilage. It acts as a scaffold that the skin can grow into, before we skin-graft. Importantly, it has no sweat glands or hair follicles, so it doesn't create any oils – it's more sterile. This reduces the bacterial count.'

By reducing the fatty bulk around the implant, they created a drier, more sterile interface. I look closer at the image, where the skin meets the silver rod of the implant, exiting the middle of the femur.

During our chat Demetrius receives a phone call and has to go and see a patient. I'm left for twenty minutes to watch patients shuffling past the café. I think about Jon and

Demetrius standing over Jack's prone body. And see them cutting through the skin and refashioning the stump, the drill coring out his femur and hammering in the implant. They must be used to it now, but I wonder if there is still a thrilling pause of awe before pushing the scalpel down and slicing through the skin. Demetrius said you need a certain amount of arrogance if you're going to cut into someone else's body and think you can leave them better than they were before.

I remember my bayonet training when I first joined up. It was a cold, wet, exhausting morning. We'd been screamed at by colour sergeants. They had ordered us to run through the drizzle to a post, faint in the distance, and back, and screamed at us to run again and again until we were sucking air in huge gulps, our mouths were dry and our heads were filled with our heartbeats, and we hated them. And they made us run again. And we ran until the effort of running meant we didn't feel. They made us run until our inhibitions had altered and then they made us scream too, and we fixed our bayonets and charged at hay-bales dressed in jackets, and stabbed them. It was an exercise in controlling aggression. It was also an exercise in understanding what it takes to pierce the surface of another person.

When Demetrius returns, we talk about alternatives to osseointegration. I ask if the future is not simply more prosthetics, but when we can grow an organic body part and reattach it. He points at my arm and says, 'I could cut your hand off with a guillotine right now – I'd take you to theatre and put it back on.' My fingers tingle when he says this. 'You'd be fine. A transplant is actually a very straightforward operation. The problem is not the operation, but the immunosuppression required when the hand comes from a donor. If you're taking drugs every day so that your body doesn't reject a transplanted hand or organ, there's a risk you may be shortening your life by, perhaps, twenty per cent. That's the trade-off. It's why we have to go through a long and involved consent process before transplant.'

Demetrius goes on to explain how reattaching a body part or organ grown from the cells of the host would overcome the challenges of transplant rejection. We can now create a vascularised chamber in a rat model and, when seeded with extracellular lignins and stem cells (those capable of regenerating and developing into specialised cells of different kinds), the granulation tissue that forms can be changed into a variety of different tissues. The stem cells can make fat, cartilage, bone. It's just the start of regenerative medicine.

'There's things we can do now, but they are simple tissues and structures,' Demetrius says. 'But a whole arm? In our lifetimes? I'm not sure.'

The hurdle we have to overcome is one of complexity. We would need to create a scaffolding that the new body part or organ can grow around – technologies like 3D printing can help with this – but we'd also need a way of making sure the correct chemical and genetic signals are in the right place of the extracellular milieu surrounding this scaffolding, so that, as a stem cell comes along, it is *told* what type of tissue it should generate at the precise time and place.

Demetrius gives an example. 'If a liver was just a huge blob of liver cells, it would be easy to regrow – we'd be doing it now. But it's a fiendishly complex structure. When the liver is damaged by alcohol, it tries to regrow, but it can't; the very specific structure and fine architecture become disordered and the organ is progressively less efficient. Our bodies find regeneration of this sort hard enough. So you could only grow a liver artificially if you could make a scaffold accurate and intricate enough, seeded with the exact information at every correct point to tell a stem cell what to do. And then, of course, you've got to keep it all alive while it grows.'

We've started walking towards the exit.

'I don't think it's crazy to suggest that regenerative medicine on the sort of scale you're talking about might be possible, Harry – the age of the transplant and prosthetic probably is finite, and we may well be able to regrow body parts one day, but it's a long way off.'

On the way home I am on the hot, sunward side of the train, and several articles lie unread on the folding tray. It had been inspiring, meeting Jon and Demetrius. They had both seemed energised by their work. It had meaning and purpose for them. Even when I'd asked them about a post-antibiotic world, they had been upbeat. Jon had admitted that most orthopaedics would be impossible without antibiotics. 'We certainly wouldn't be doing osseo,' he'd said, 'but something will turn up. When it does, it will completely revolutionise surgery.'

I run my hands over the hard, smooth surface of my sockets; they are uncomfortable and hot after a day of travel, but I can't remove them while squashed into the seat. I wouldn't anyway – the smell of the liners would be unpleasant for the woman sitting next to me. I'll live with the pain till I get home. I wonder if there is osseointegration in my future. Jon had said it was still last-ditch. 'If we can close the abutment where the skin meets the implant and stop deep infection, then maybe it will be offered more widely, but we're not there yet.'

The sunlight streaks a web in the scratches on the carriage window. I see Daedalus attaching wings to Icarus with osseointegration. Jack is flying close to the sun. But at least he's having a chance to fly.

Freedom Is Expensive

The Karlsruhe Trade Fair Centre consists of four 12,500-square-metre aircraft hangar-like structures connected by a façade of conference rooms, atriums and events halls, all glass walls and sterile and ready for the 300 various annual fairs, conferences, concerts, tech shows, seminars and symposiums that make use of its location, pretty much in the middle of Western Europe, not far from the France–Germany border. I'm standing in the central concourse that runs along the spine of the complex, amid coffee shops and beige-food outlets not yet open, looking at the list of exhibitors. I'm here for the biennial REHAB European trade fair – strapline: *Rehabilitation | Therapy | Care | Inclusion*; slogan: *The Trade Fair for Improved Quality of Life*. The website says there are 460 exhibitors, from twenty-one countries, and I will be one of about 18,500 visitors, 35 per cent of whom will be disabled or carers, while the rest will be specialist visitors, which I suppose means they work in the med-tech industry. The map I was handed on entry is in German and shows a sea of logos in boxes. There's a thematic layout, but none of it means much to me, so I head into *Halle 1*.

Below the vast curved roof and the hanging lighting gantries strung with logos and signs is the gridded mini-city of exhibitors' stands. I make my way down the first avenue, planning to do a quick circuit and get my bearings. Each stand is divided by temporary walls headed with logos and straplines. There are various sizes: some stands, for the larger companies, are imposing, with steps and stages and display cases with LED-lit matt grey devices inside – as if they have shipped a whole high-street shop and fitters to the fair; others, tucked away in the corners of the hall, are smaller market-like stalls.

I walk, pulling my wheelie case (I'm just off the flight). The hall is almost empty, the fair has only just opened its doors and some of the stands are still setting up – there's a woman in an orange T-shirt on a stepladder tying a bunch of orange-logoed balloons to the top of her step-and-repeat backdrop. I walk past some of the bigger stands – an area of various high-end wheelchair companies; a couple of the big prosthetic companies I know; and another company that sells portable and ceiling-mounted patient hoists, next to one with what look like huge steam ovens for sterilising rehab aids in care homes – and head into an area of smaller stands. A unifier of almost all the stands seems to be one of those pod coffee machines, with a stack of cardboard coffee cups. I've already been offered two as I've lingered to look. They seem to have a dual purpose: to entice potential customers, and to get the exhibitors through the next few days.

I'm not sure which thematic area I'm in now, but I have a feeling it's less expensive real estate. The stands are small, less branded. And it's not so business-like here, more jovial, and the stallholders can't slip into English as easily as the reps up near the entrance. Here there are herbal and homeopathic remedies; cloth-sewn neck-warmers, which I think go in the microwave; decorative personalised walking sticks; and accessible barge holidays. I stop and look at the model of the barge with a little doll in a wheelchair – another coffee offered, and leaflets pressed on me. Further on there's a stall selling a variety of children's puppets; a wall of fuzzy colour, with rehab and additional-needs literature on posters; and another that seems to be selling rubber overshoes. I get the impression many exhibitors are independent inventors or retirees with a little more time on their hands; they've brought their good ideas to market to shift some stock, or perhaps to get spotted by one of the giants who takes an interest – a buyout might follow.

Disability and invention have always gone hand-in-hand. There are examples that have changed the face of the prosthetics industry: James E. Hanger lost his leg to a cannonball early in the American Civil War. He returned home with the standard-issue peg-leg, which he hated, and knew (he had a background in engineering) there must be a better way. He spent the next ten years developing a prosthetic leg with hinges at the knee and ankle, and rubber buffers to keep the thing from making too much noise. By the end of the war there were around 45,000 amputees, and Hanger had factories manufacturing his legs for them. Now Hanger, Inc. is a multinational prosthetic and orthotics company, with 4,900 employees and a 20 per cent share of the US market, which is worth approximately $4.2 billion.

Physical impairment creates very practical problems, and if you're the one living with the problem, you tend to spend a lot of your time working out how to overcome it. We all imagine a better life – for most of us, it's aspirational daydreams or the lottery-win *what-ifs* – but with a disability, solutions that will improve quality of life can be very concrete. And often there seems to be a blatant gap in the market. As ever with successful inventions, it's a marriage of technical knowledge, determination and an idea that has a wider benefit than for solely the specific individual who has been tinkering away in the garage. There are many examples.

I have a Bartlett Tendon Knee in a cupboard at home. Brian Bartlett was a member of the US Ski Team when he lost his leg in a traffic accident. Usual story: he was told he'd never ski again, so he invented a prosthetic knee that could support the user's body weight during the high impact of extreme sports. The idea is ingenious. He used external rubber tendons that wrap over circular rollers at the knee joint, mimicking ligaments and muscles and letting an amputee ski or mountain-bike. Bartlett went on to partner with a prosthetic company and brought his idea to market.

In 1981 Dan Everard created the Yellow Peril, a little electric wheelchair (it looks like a miniature yellow forklift: big wheels in the front, castors at the back and a seat that can elevate or lower to the ground) for his daughter Ruth, who was diagnosed with muscular dystrophy. She started using the chair just before she was two. It gave her the independence to have almost all the crucial non-verbal developmental experiences that children only get through mobility – touching the radiator, for instance, and understanding it's hot. It led to the not-for-profit Dragonmobility Ltd, which now makes elevating wheelchairs for all ages.

I know some of these inventors. A friend of mine, who was an engineer in the military and lost both legs in Afghanistan at a similar time to me, now has a PhD and works at Imperial College London, designing (among other things) a smart socket that can tell the user if their stump is too hot or under too much pressure; a leg in a box, for low- and middle-income countries; and an internal prosthetic knee that can be implanted into an amputee after injury.

And the man who fitted me for my first legs retrained as a prosthetist when he lost his foot in a motorcycle accident. All these people were taking a little control over their situation, trying to solve the problems that disability created in their daily lives.

I browse into an area of the fair where scores of mobility scooters are displayed on a huge round plinth, and beyond are all kinds of adapted and accessible vehicles: cars with clever driving aids and cantilever wheelchair lifts; vans with electronic ramps; and even a motorhome with a smaller runaround convertible car stowed inside its chassis. Next is a zone where all the stands are themed around virtual reality (VR). Two people are in a booth, sitting in chairs with headsets on and holding a games controller-type device and trying to manipulate something in whatever world they are in. One grunts in frustration, the other is smiling. They seem to be playing against each other. Ahead is a short queue; people are waiting to have a go on an exercise machine in which you VR-fly while lying on a tilting frame. I join the back – there's a teenager on the machine at the moment, his VR goggles peering around blindly. His shorts keep falling down, and his embarrassed mother tells his laughing brother to reach in and pull them up.

Virtual reality is now used across medicine. Multisensory and three-dimensional simulated environments become a place where young doctors and healthcare professionals can train without being a burden on the healthcare system. In the past, trainee surgeons needed time-consuming hands-on experience under the supervision of senior surgeons. But in VR environments their performance can be recorded, data collected and compared, and new methods and instruments such as robotics can be trained on; and, critically, trainees learn by doing *before* actually doing. There are disadvantages: VR is initially expensive, it reduces real human connection and it's consequence-free, so there is always the risk that medics might carry the cavalier approaches they have learnt virtually into the real world. But for repetitive diagnostic procedures such as colonoscopies and endoscopies, for learning anatomy and for mocking up real-world emergency situations for paramedics, and virtual sessions for GPs, it's a powerful tool.

VR has clear benefits in rehabilitation, where recovery is often goal-orientated and requires people to be motivated and perform tedious and repetitive movements. Making people stick with their programme of exercises is difficult, especially when they leave the watchful eye of a therapist (I always lied to my physio about how much I did). VR games that encourage the patient to perform movements are becoming more immersive. Playing a VR game with the challenge of completing increasingly difficult tasks by twisting and moving a joystick or wand, or through whole-body movements in front of motion-capture programs, strengthens and increases range, improves function and keeps patients interested.

When I get to the front of the queue, I'm given some brief instructions from the marketing rep (tight, branded T-shirt tucked into chinos) and climb onto the rig. I'm sort of lying spreadeagled on it, with my legs in stirrups behind me and holding two handles in front of me – I suppose in the position you'd be in if you were holding on to the neck of a giant bird in flight. The whole thing is very unstable and pivots and rocks on a number of different sliding rollers. I find a shaky balance and lower the VR headset over my eyes. I'm high above alpine mountains in the summer, with pine forests and lakes below, flying towards a glowing green ring, with a score hovering ahead. It's approaching quickly now and I'm not going to make it through, and I over-correct and the whole thing wobbles and clunks and I veer widely up into the sky. I find the centre again and can feel my core engaging and beginning to ache. I fly through a ring, just make it over a rocky ridge and dive onto a line of rings in the valley ahead. My score counts up. My core is really burning now. I also have a feeling that one of the reps is steadying the machine to give me a hand. After flying down a valley, trying to get through as many rings as possible, I clamber down and thank him. I'd used rudimentary VR rehab programs before – slab graphics and simple tasks, stacking coloured blocks or squeezing and twisting a joystick to make a car turn. This was different. Not only were my abs aching (and would be for the next few days), but it was fun. Escapism can be hard to find if you're disabled. It did feel, for a moment, like I'd left *Halle 1*.

I exit into the open air, cross an asphalt courtyard where a couple of people are testing out three-wheeled electric trikes and enter *Halle 2*. More of the same – every imaginable disability aid. I pass a stall of attachments that motorise wheelchairs: a large off-road tyre and handlebars, which you bolt on the front of the chair, is the flagship product. Behind this is a queue of people waiting to try out a futuristic wheelchair. I watch one pink-haired woman waiting in her battered, boxy electric wheelchair, covered in rock-concert stickers with square battery and exposed wires all slung underneath. Her friend helps her transfer across. The old chair was an extension of her goth style and she looks out of place in this shiny machine – it's two-wheeled and gyroscopically stabilised. There's something magic in the way it stays upright (a bit like a Segway but stranger, because the centre of gravity is so high), and it wouldn't be out of place gliding into a scene of *Star Trek*. The marketing rep (branded hoodie this time) shows the woman the controls and she grasps the control stick and moves the chair out into the crowds. The machine leans her forward and accelerates with all the smoothness and whirring of new electric vehicles, and the rep jogs after her, one hand out, ready to grab the top of the seat if she gets it wrong.

There are bar stools around a pod with questionnaires, and I take a seat to watch. As the woman returns, her pink fringe flutters on her forehead. She is genuinely beaming. The rep shows her another feature, and a pair of caterpillar tracks lower from the chair's undercarriage and it climbs backwards up a short flight of demonstration steps – it seems stable enough, but the woman is nervous; it does look precarious, but the rep gives the machine a push to demonstrate how solid it is, which makes her shriek. When she descends, the rep shows her another function and the chair deploys two pilot wheels behind, which – in tandem with the caterpillar tracks – create a lifting triangle, and she rises until she is at the same height as her friend and they are both laughing.

In my ten-year experience of disability there have been a handful of things that have been truly life-changing: my microprocessor knee; learning to drive again; my kayak; and my electric Swifty scooter. These are all items that have given me back my freedom. Not only that, but they were all paid for by the NHS, or applied for through charities. I can't really ride a bike any more (the Bartlett Tendon Knee was okay for a bit, but, with two legs missing, it just wasn't practical), but Swifty means I have personal electric mobility, can keep up with my daughter on her bike and can commute without relying on a car. It is wind-in-the-hair freedom. Personal electric mobility seems to be here to stay (for everyone, not just the disabled), with scooter rental companies already changing the environment of big cities, and more commuters opting for them.

The woman is back in her own chair now. It suits her better; there's something about the sleek grey and LED lights of the new wheelchair that is slightly over the top. She is talking to the rep. I imagine they are discussing cost and timelines. (I look them up later: around €30,000 – which, given the price of my prosthetic legs, seems good value – and a year-long waiting list.) Then her friend is writing on a clipboard, presumably the woman's email address. I wonder what sacrifices she'll have to make to afford it. And I suppose she'll choose black and turn the lights off, and cover it in stickers, and it will suit her too.

The next day I return to the fair. It's noisier. If Friday was for the marketing-rep catch-ups and inter-industry conversations, today (Saturday) is for the people. All the food and drink outlets are open, and the place is packed. I pass a game of wheelchair rugby in the foyer and join the sea of atypical bodies propelled by every kind of mobility device, prosthetics and orthotics, and parents pushing their children with multiple severe disabilities, flowing between stands, all searching for the latest gadget that will make their lives a little easier. I notice people are carrying red branded buckets, using them as receptacles for marketing material handed out at stands and by students clad in the branded liveries – pens and pencil cases, leaflets and brochures. And keyrings are popular; there's also a mug and a small soft toy that seem particularly prized by two young girls comparing their hauls.

A few World Health Organization facts: worldwide about fifteen people in every 100 have a disability, and between two and four people in every 100 are severely disabled; there are 500,000 spinal injuries each year, and sixty-five million wheelchair users (and between ten and twenty million people need a wheelchair but cannot afford one); in the next thirty years the number of people over the age of eighty is set to almost quadruple, to 395 million; and in Europe there are currently forty million people who cannot walk without some sort of mobility aid. It's a large market.

The European medical technology industry (devices, diagnostics and digital products, or solutions used to save and improve people's lives) represents a €120 billion market. It files around 14,000 new patents a year, the second highest of all sectors behind digital communications. It employs 730,000 people – Germany has the most, at 227,000 (twenty-eight per 100,000), making Karlsruhe a good home for this trade show (although Ireland has the most per capita, at eighty-three per 100,000). It's an industry that has an important economic and societal impact and it is, of course, a competitive market.

The REHAB fair isn't as garish as some marketing environments, not so aggressive, but it still plays on our aspirations. There are posters of beautiful people with their mobility aids in beautiful settings under slogans: *It's time for boundless freedom*; *LIFE WITHOUT LIMITATIONS*; *It's not just walking – it's More Than Walking*. The market forces of capitalism are in play, and they are after us: the limping, wheeling, blind, deaf customers who have come to find out which of the latest products will make our lives easier. As you'd expect, some of this tech is mind-blowingly expensive – someone has to pay for all those clever scientists and all this branding.

And there's inequality here too, of course. Some of the people moving around the maze of stalls will have had insurance payments of millions of euros – the motorcyclist knocked off a bike; the worker caught in an industrial accident; or the child starved of oxygen by a negligent hospital. Many will have fought through the courts, waiting years for payouts. But even millions of euros only go so far, when spread over a lifetime. Very few of the people at this fair won't have to weigh up carefully whether the benefit of a new assistive technology is worth the outlay; whether to wait for the next generation of device or make do with what they've already got. Then there are the unknowns, hidden behind the slogans. What if it doesn't quite do what it promises or needs replacing soon; how long is the warranty; and what if my very specific disability isn't quite catered for? But what price do you put on being able to walk, or get upstairs in your home, or read the paper, or drive a car? Imagine that you had never been able to go to the beach, or for a walk through the woods, but a new wheelchair might offer you the chance.

I'm sure there will be some people who haven't even shown up, even though they have dreamt of it. They are the ones who slip through the gaps and have little support, left to struggle on because their disability was congenital, or misunderstood, or no one was to blame for their accident. Add other structural inequalities – poverty, age, race, being a woman, being unemployed, doing badly at school – which all mean you're more likely to be disabled. And there are those too proud, or unable, to ask for help from charities and support groups; they are the people who will never be able to afford some of this kit.

This huge fair is a shop front. It also represents the fact that there's a hidden tax on being disabled. Being a hybrid human means expensive kit – you have to pay for the privilege of leading a normal life.

Ahead is the Ottobock stand. It's one of the big ones and has the feel of a luxury sports shop. It's the company that makes my bionic prosthetic knee, among many other medtech devices and services, and is valued at $3.5 billion – the Apple or Microsoft of prosthetics companies. There it is, uplit on a plinth, one of their flagship products, with the poster behind: *Reclaim all you want to be*. The strapline trying to sell a dream – and I know, even though it is the best prosthetic knee out there, reclaiming all I want to be is an un-keepable promise, tinged with a little of the hollowness of so much branding. But *Reclaim all you want to be within the current constraints of technology, and at a practical price point* wouldn't have flown in the marketing meeting.

And yet, for all the inequality and spin, I have the feeling there's goodwill and altruism behind the transactions going on here. So many of the companies represented are not-for-profits, and even in the big stalls like this one, where the marketing is slick and flashy, there's more than simply hard-nosed business. So many of the scientists I've met described the personal motivation for what they were doing – a disabled relative or childhood friend – and it's the same among the reps and company management. Fundamentally, it's about making people's lives better. Maybe it is about as noble an industry as you can get.

Across the stage, two models are chatting. They have seen me looking at my knee on display and wave and give a thumbs-up. They are tall and good-looking. Her prosthetic leg is remarkably eye-catching, between the high-heeled ankle boots and black leather miniskirt. He is a muscled hipster with a trimmed beard, shaved head and wearing shorts, and there is a certain sexiness to the prosthetic alongside his tattooed leg. I smile and wave back.

I'm not suggesting that everyone is going to want to head to their next national disability and assistive-tech fair – it's no Geneva Motor Show – but there is some amazing kit on display, and it's kit we're all far more likely to use in our lifetimes than a supercar (by some estimates, we're all likely to spend around 20 per cent of our lifetimes with a disability). It's not much of a leap to think that the exoskeleton I'm now standing in front of, at the next stand, will be the mobility scooter of the next 100 years. And perhaps if more of us knew about the technology on offer we might be better at adopting it when our bodies start to deteriorate and need assistance.

The main reason to I travelled to Karlsruhe is taking place in a football pitch-sized area in the corner of *Halle 2*. As I approach I can see through the balloons and stand logos a huge suspended LCD screen playing a slick montage, and I start to notice looping, propulsive Hans Zimmer-esque music over the noise of the crowds. The Cybathlon is a sort of mash-up of Formula One, the Paralympics and a series of elaborate party games. It was set up by Robert Riener, a Swiss professor of sensory-motor systems, to break down barriers between the disabled, the public and the assistive-technology developers, with the goal of better integrating disabled people into society. It's essentially a sports day – a hybrid Olympics – where med-tech companies and university bioengineering departments can gather to test their latest ideas and devices against each other. The competition is fun and creates a spectacle, but it's mostly about collaboration, exposure, research and, critically, ensuring that people with disability are in proper dialogue with those who develop the assistive devices they use.

The first Cybathlon took place in 2016 in Switzerland in an adapted athletics arena. The packed-out audience watched sixty-six teams from twenty-five countries take part in the six different disciplines: exoskeleton race, arm-prosthesis race, leg-prosthesis race, brain–computer interface race, powered wheelchair race and functional electrical-stimulation bike race. Here in Karlsruhe we're between full Cybathlons (like the Olympics, it's a four-yearly event), so only the arm-prosthesis and leg-prosthesis races are being showcased. It's a sort of travelling qualifying series, set up to promote the event and drive scientific exchange between teams – this time making use of the REHAB trade show as a venue.

Right now the leg-prosthesis race is about to start, and

the two pilots (the term used for the disabled competitors) are being interviewed at the start of the track. A presenter holds a microphone up to them and asks questions, and it's beamed onto the huge LCD screen above us. There are a few people watching, having drifted over from the stands to see what's happening, sipping beers in plastic cups and eating huge pretzels. Then the presenter and cameraman are out of the way, and a graphic fizzes across the screen, counting the pilots down with a percussive heartbeat sound effect: *3-2-1 GO*. They race down two parallel tracks with a series of obstacles in the way. It's not exactly high-octane. Before they can get up a head of steam, they have to sit and stand, holding a plate and cutlery; then they step over a series of hurdles balancing two apples on plates – the nearest pilot drops one and it rolls away and he's back to retry the obstacle; then it's carrying a crate and ball over some steps; after that it's across a cambered slope, carrying more apples. Green flags are lifted, with that extremely snappy action they use in all sports, every time they complete an obstacle. A red flag goes up if the pilots make an infringement or rely on their good legs when they should be using their prosthetic ones. One of the pilots has the same leg as me, while the other has an Össur Power Knee. It's the Real Madrid v. Barcelona of prosthetic races.

I lean on the blue-logoed hoardings – *CYBATHLON | MOVING PEOPLE AND TECHNOLOGY* – that run the length of the 50-metre or so course and watch them pass. At first glance (and I imagine to most of the able-bodied spectators) it looks like a glorified egg-and-spoon race and lacks the athletic speed, grace and gritty competitiveness of the kind we're used to in good sports entertainment. But as someone who uses a prosthetic leg, I can see how difficult the tasks would be: stepping a prosthetic through the hurdles while at the same time balancing apples on a plate; walking along a cambered slope; standing from a chair without pushing up with your hands – all very hard to do quickly and without dropping what you're carrying. I can see the determination and human skill.

But if you don't know this, or don't look hard enough for it, it all seems a bit laboured, and you'd be sipping your beer and eating your pretzel wondering when something exciting was going to happen. No one cheers at the sublime, imperceptible movement the pilot has just made in their thigh to engage the algorithm that makes their prosthetic step over a hurdle, all while balancing on the other leg and holding a plate with two plastic apples – it goes unnoticed. But I can see it, and I know that I'd have fallen over and the apples would be on the floor; I know it is a skill-move comparable to an overhead kick, or a backhand topspin lob, or a fourth-down no-look pass.

I'd heard Robert Riener speak. He said he set up the Cybathlon for lots of very marketable reasons, but he had other less obvious ambitions: showing what our taxes and research-and-development grants pay for; trying to help med-tech companies bring their product to market; and, one of the most important reasons for him, showing people the reality: that we're a long way from the dreams of science fiction. People's expectations are distorted by popular culture.

When you see Iron Man or the Terminator or Luke Sky-walker in the movies, you start to think prosthetic devices can outperform our human biology. I'm often stopped by people in the street and asked how high I can jump, or if I can run faster now I have prosthetics – and I have to tell them I can't jump or run at all in these legs, and they look confused and disappointed, and I feel I've let them down. And watching the two pilots painstakingly make their way down the course is underwhelming to the untrained eye – these are two of the best microprocessor knees around, but they still have a very long way to go before they equal the performance of the average human limb, let alone the fantasies of science fiction. But look closer, understand a few of the developments that have been required to get even this far, and you see that the technology that is letting the pilots do this is astonishing; it's come a long way already.

The race is over, and up on the big screen the winner is being interviewed and I walk towards one of the food outlets. I grab a beer and sit near the start. On the course they are changing over the tasks, ready for the upper-limb race, and there's a lull. A man starts limbering up next to me, like a 100-metre runner, rolling his shoulder and doing little verti-cal jumps from his ankles. He seems nervous.

'Good luck,' I say.

He's called Bert and is one of the pilots up next. He has a Touch Bionics i-limb. Bert is Dutch and tells me he's been practising in his garage.

'I'd never unscrew a light bulb with my prosthetic – I'd just use my real hand – so I've been practising, you know.'

I ask him what it's like being an ambassador. He tells me how great it was being at the first Cybathlon in 2016 – life-changing; the travel can be good too, and he likes trying out the latest kit and software upgrades: it makes him feel useful, and part of it.

'You should try it,' he says.

We talk some more, then he is called over to the start and is enlarged, up on the big screen, being interviewed.

Much like the leg race, the arm race is a series of tasks – and for the prosthetic hands the tasks are far more intricate. First is a table with different objects. The pilots have to cut a loaf of bread (foam), unwrap a chocolate, open a bottle, jam-jar and tin can, then take a match out of a box and light a candle; then it's on to a washing line, where they have to put on and zip up a hoodie, then do up two buttons on a shirt. Bert is struggling with the buttons, and I notice the other pilot is using her teeth to help, which surely isn't allowed. Bert said there was a risk for companies putting their products on display like this, testing them against the competition in front of the public. It shows their limitations. I hadn't thought about it like that – I suppose, if I was in the market for a new prosthetic, I might come and watch how they performed against each other.

Bert is catching up on the laces task: he has to tie two shoes together and then hang them over the washing line. It is amazing watching the bionic fingers pinch the lace. Then he has to put a variety of objects into holes: a USB stick, a key. He's now concentrating very hard, crouched down, making the fingers close onto a credit card, which he has slid to the edge of a desk. There's something slightly painful about watching someone struggling to pick up a credit card when you know they could simply use their biological hand in a single motion and push it into the slot. Bert's getting flustered now, obviously trying to tense the muscles in his stump next to sensors that activate the prosthetic fingers to close slowly around the card, and the seconds are counting up. It's a reminder of why, because of the training and cognitive effort of operating them, around 50 per cent of people abandon these bionic hands.

He's on the hammer-and-nail task now, then manages to screw in the light bulb before cutting a sheet of paper with movements of the limb that seem almost magically human – he's into his stride again. Then there are six boxes. He has to reach in and identify by 'feeling' with his prosthetic (and no visual cues) what is inside, before placing six corresponding objects on top of the boxes: a ball, a block, a piece of foam, and so on. I notice the other pilot is banging the object against the inside of the box to get an audible clue – again surely not in the spirit. There are a few cheers at the green flags each time the pilots get one right. And then there is one of those ring-and-buzz wire games.

I'm not sure if Bert won or not; there were green and red flags going up on both sides of the course. It doesn't seem to matter. When he is interviewed afterwards, I'm surprised to see him crying. He can hardly answer the questions, he is so choked up. The camera zooms in so that *Halle 2* has a close-up of the emotion on the big screen. It obviously meant a lot to him.

Later, at the prize-giving, Bert punches both biological and bionic hands into the air in triumph, as he jumps onto the third step of the podium. He kisses his medal; it's as if he has won the Olympics, and there are a few whoops from those still watching.

Raging against the Dying of the Light

A few weeks after returning from Germany I meet Jamie for a walk in the park. I see him in the distance, waiting to cross the road. The lights change twice and the cars stop and go, stop and go, and still he doesn't cross. I want to get there to help him. Then the barman from the pub on the corner is rushing out to guide him.

'They've taken the beeper out of that crossing,' Jamie tells me as we enter the park. 'It's a real nuisance.'

We begin walking, Jamie's white cane skating across the path. There's a slight ridge in the centre of the asphalt I imagine he is feeling for. I got to know Jamie soon after I lost my legs; he lives close by. He'd always told me how being blind let him in on the secret of human kindness: 'People really are bloody kind. That's what I see – the quiet decency of the British public, who unfailingly offer a helping hand.'

The barman had helped him across, but I'd seen thirty-odd

people walk straight past him. Jamie also gets animated about that. 'It's the terrorism of political correctness, Harry. People worry they are going to offend me if they ask if I need help. Yes, I don't actually need help for steps or a kerb – they're predictable and I can see them with my white cane – but I do need help if I'm about to walk into a huge wire-cage dustbin that's been parked in the middle of my normal route. That happened, you know. I was covered in blood, and a dog-walker threatened to tie me up with his lead if I didn't wait for the ambulance. It doesn't matter if people use the wrong words or offer help when I don't need it. I know they're only trying to be kind – and I'm grateful for the thought. It's those disabled people who, when asked if they are okay, reply, "Yes, are *YOU* okay?" as if it's some sort of affront. Can't they just be thankful that someone is trying to help?'

It was a reminder that I'd been both those people: replying to offers of help with a gracious, 'I'm okay, thank you', but also prone to a prickly 'NO, thank-you', charged with an unspoken *Mind your own fucking business.*

Jamie has retinitis pigmentosa, a congenital disease where the cells in the retina at the back of the eye break down and the visual field progressively reduces. When he was twelve he was told that he would be completely blind by nineteen, but it actually took until he was in his forties. 'I've never seen the stars,' he told me once. 'As a child I was night-blind and had to feel my way into the house from the garden when all the other children ran back in.'

Hearing this, I felt a surge of pity for him, I suppose because I value sight above all other senses. When I was blown up, my sight was the first thing I checked: *Yes, I can still see. That's the most important thing.* But in the minutes that followed, either through massive blood loss or conversion disorder – what they used to call 'hysterical blindness' – everything went dark and I was blind. It swelled the terror and loneliness of those minutes, and I couldn't help but place those emotions on Jamie. But he instantly snapped me out of the pity.

'Hey, Harry, it's not like that. Happiness is about accepting your fate. Pity is ridiculous. The vibe, the warmth, the laughter, the kindness – all these have grown for me. You know that; you've told me the same.'

I've talked to Jamie about this before, about pity: that sometimes people think the disability is the largest part of you, that it must infect all of who you are, and you must be challenged in every way. 'They are surprised if you make a funny speech at a wedding or write a letter to the paper,' he said. 'When actually disability is hardly any part of who you are.'

Jamie has always seemed upbeat and full of gratitude, and he can be very funny about the absurdity of it. 'Because disability is absurd, isn't it?' he said. 'Like the time I was in a hotel and cursing the staff for putting excess garnish in my orange juice, until I heard the waiter whisper in my ear, "That's the vase for the rose, sir."'

People call out to Jamie. He's here every day, no matter the weather, and is part of the park's character. He stops and chats, two hands resting over the top of his cane, like a sage with his staff, searching beyond the horizon. As we walk on and I watch the white ball-tip rolling back and forth across the tarmac, I find I am telling him enthusiastically about a few of the devices I'd seen at the REHAB trade fair in Germany. A small camera device mounted on glasses that can read for you. You look at the page and it starts speaking the words; it can be trained to identify faces and tell you what you've picked up in the shop, by reading the barcode; and the colour of things; and the value of the banknote you're holding; and, if you twist your wrist up to your face, what time it is. And now I'm telling Jamie about a technology-assisted white cane with an ultrasonic detector that warns of unexpected obstacles in your path, like a low branch, by vibrating in the handle. It can link up to a smartphone and act as a navigation aid by speaking directions.

'You have been busy, haven't you?' He's smiling as he walks. 'Yes, I'm aware of this stuff, Harry. I get the newsletters and articles forwarded on by my supportive friends and family. I'm happy with what I've got. And they're heavy, and you have to plug them in, and this cane works for me just fine.'

Jamie didn't want them. It wasn't that he didn't want them in an entrenched, dogmatic way, but simply that he didn't feel he needed them. He was happy with what he had. He also told me that any device that vibrated or beeped or, worse, spoke would interfere with his own senses, which are finely tuned to his environment – his natural hearing, his biological feel and proprioception on the cane are far more sensitive than any technology. His experience of being blind was *visceral* and *elemental*; that's the way he described how he navigated the world. So many of these high-tech devices intruded on that.

He stops and rests his hands on the top of his cane again. 'I'd rather embrace my life the way it is, Harry. I think it's best to *wear* a disability. It's not an enemy to be fought, it's like your attitude to life: if you constantly fight against it, it only leads to unhappiness. I don't want to keep looking for how to make it better. I'm happy with how it is.'

We're walking again and Jamie turns the corner down another path and I'm not sure how he knew to make the turn when he did.

'You won't find me raging against the dying of the light,'* he says. 'Blindness has enriched my life. I don't want to spend the rest of my life clinging to some distant hope it can be fixed – it's always disappointing.'

We've finished a loop of the park. Jamie's going to keep going, but my legs are spent. 'You know where you are, Jamie?'

'Yes, all good, thanks, mate. See you soon.'

When the first iPhone was unveiled by Steve Jobs at the 2007 Macworld trade show, the blind community panicked. The world was being shown the interface of the future and it was a flat screen. The visually impaired were going to be *left in the dark* without the physicality and feedback of buttons. But Apple was pioneering accessibility in its devices and wanted them to be as easy as possible to use – for anyone, no matter what their impairment. It quickly released VoiceOver

---

*A refrain from Dylan Thomas's poem 'Do not go gentle into that good night'.

screen-reader software, which spoke an audio description of the icon or text that the user's finger was resting on; and a zoom function that could magnify the screen up to twenty times; and options to change text and cursor size and invert screen colours, providing countless options for legibility: useful for all kinds of disabilities and diseases, from dyslexia to migraine sufferers. Today, many of the features designed to make Apple products intuitive, and as adoptable as possible for everyone, are invaluable for the disabled: voice-command digital assistant Siri can open applications or change settings, send messages, start calls and search the Web; dictation software converts spoken word into text; and you can send an email or message without having to tap the flat touchscreen keyboard (good for the visually impaired, but also for anyone with arthritis or motor impairment). And the deaf can now use sign language on video calls; and for those with the most severe motor impairment there is *switch control*, where everything can be controlled by pushing a single button or joystick. Developing new ways of interfacing, making accessibility central to the design of their products, led the president of the National Federation of the Blind to say, 'Apple has done more for accessibility than any other company to date.'

Mobile electronic devices may well be the most important technology advance ever for the visually impaired. The explosion of smartphone and tablet-computer use in the general population has been mirrored in the blind community. Smartphones can be used for reading, writing, listening to audiobooks, communication, navigation and object identification – no wonder they are replacing as many as eleven different, often bulky, expensive stand-alone devices the visually impaired traditionally used. As much as 30–50 per cent of all traditional assistive devices are thought to have been abandoned by the visually impaired; smartphones are often cheaper, easier to train on and, critically, don't have the stigma of being assistive devices and signalling to the rest of the world that you have a disability.

So much of our tech is dedicated to solving one problem. The running blade, for instance, is brilliant at propelling the amputee along the running track – in some cases it might equal our biology – but is next to useless at any other task (making a cup of tea while on blades would be like trying to do so while tightrope-walking). The smartphone can perform multiple assistive tasks and live in our pockets; it's the reason nearly all of us find it so compelling.

When Hayden returns to his apartment in New York City he uses Microsoft's Seeing AI on his iPhone to check the post. He holds each letter up to the phone's camera and the app instantly reads aloud to him – even handwritten notes. He can sift his junk mail in a matter of minutes. The alternative is turning on his desktop computer, setting up the scanner, lining up the documents just right, hitting the scan button and waiting for it to process. The Seeing AI app collapses the length and complexity of the task. It can also be used to read product barcodes, recognise friends, describe a scene around him and identify currency. It is one of many free or small-subscription apps Hayden has downloaded on his iPhone.

'I use a number of apps to get around the city,' he tells me when he speaks to me from New York. 'Ride-hailing apps, like everyone does, and apps like Seeing Eye GPS for navigating, which can direct me, but also tell me what landmarks and shops I'm passing. I can ask it to take me to the hardware store, even though I've never been to the hardware store before. In the past I'd need someone to guide me. I'm using Google Maps more now; it doesn't have such detailed directions for a blind person, but it can give me all sorts of useful information. If I'm going to a restaurant, I can look it up in advance, pull up the menu and figure out what dish I want to order, even before I get there – I don't have to rely on my friends to read the entire menu to me. It can tell me when the next bus or train is. There's a whole range of possibilities.'

Hayden went fully blind just before university. He was due to study environmental engineering, so had to adapt quickly and creatively. The course was heavy on visual equations and diagrams. With the help of his professors and classmates, he made 3D-printed models that could be snapped together by assistants. They represented the electrical circuits and fluid dynamic-systems diagrams that were up on the whiteboard or in exam papers. He could interact with the models by feeling over the different printed elements and moving them around, analysing a tactile interactive version of what everyone else was looking at in the classroom. Since then he has experimented with 3D-printed graphs and converting datasets into sounds. 'I couldn't interpret the trends in huge spreadsheets, so feeling a three-dimensional printed representation of a graph, or hearing a changing tone that symbolised a graph's parabola, helped me understand the effect different parameters I introduced to the data were having.'

'Fathom is almost a prosthetic, Harry.' He laughs when he says this about his guide dog. 'I'm part dog, part human – a *dogborg*!' We are talking about dependency on technology and what it would mean if he lost any of his assistive devices. 'It would be devastating to lose my phone. It would be almost the worst-imaginable experience. I'd manage, but if I was out in the city and was pickpocketed or if it fell on the train tracks, I'd find it very challenging. But with Fathom, it's more emotional. It's like part of my memory is embedded in him; he knows my favourite entrances and benches, he even remembered routes in London when we returned after two years away. If I was partnered up with a different guide dog, those memories wouldn't be there. We have a shared memory and experience. Sometimes Fathom jumps up and down when we pass a bench – he takes so much pleasure in finding these things for me, and even though I don't need a rest, I sit to make him happy. He gives me a certain level of confidence; it's social, emotional and physical confidence. A smartphone can't do that, of course. I can't imagine life without Fathom.'

Hayden describes the developments in the five years since he left university as 'pretty spectacular'. Something very simple, like the increase of memory on the iPhone, enables him to carry multiple accessibility apps and a whole library of audiobooks in his pocket wherever he goes. New websites are now expected to be accessible – one of his greatest frustrations is reaching a website where the links and buttons aren't properly labelled (so voiceover software doesn't work) and he has to click randomly, in the hope that he'll eventually navigate to the right place. He is also incredibly grateful for AD (audio description of the action going on between the dialogue in a film or TV show). He used to rely on friends to describe important visual moments or simply listened to the dialogue, but with AD he has access to the meaning in characters' faces that is unspoken, or to the purely visual action and tension that carry so much of a story's meaning. Now nearly all big releases have AD and it makes him feel more culturally relevant.

'Siri was created for everyone, but it's probably of greater benefit to me than a sighted person,' Hayden says. 'I can now pick up my mum's or girlfriend's phone and use it. That seems to be the most important thing: having universal design built into all systems, whether it's a building's entranceway or an electronic device. That's true accessibility and there have been massive improvements. I'm very hopeful. There's never been a better time to be blind.'

Hayden and Jamie are from different generations, and I notice a different approach to adopting tech that reflects this; but Hayden, like Jamie, is suspicious of high-tech solutions for people with disabilities, especially when they are being created by people who are not disabled.

'There's a lot of hype,' he says. 'For my condition – just in terms of the pure biology – there's probably no cure in the near to mid-term. Something might eventually be possible. They're doing things in mice: wafers impregnated in stem cells and implanted behind the retina; we'll see what happens with that. But any stem-cell-based intervention is probably a long way off, if it ever happens. So I'm not holding out much hope on that.'

Hayden uses a smartphone installed with a range of apps and a guide dog; Jamie chooses not to, but I found the way both of them balanced acceptance and adaption inspiring. I still had to fight the urge to phone Jamie and tell him about the apps Hayden was using. I knew I'd be making the same mistake as I had in the park – presuming that Jamie would want, or need, any of it. Even if there was a chance it might make his life a little easier, it was not who Jamie wanted to be.

Later in the year a friend sends me a link. I get them every now and again. *Check this out, mate, it might help. Isn't this amazing? Saw this and thought of you.* Normally they are completely unsuitable or experimental, or might become reality in ten years' time – just news hype. (It makes me wonder how a terminal-cancer patient must feel when they see similar news of some genetic and highly experimental miracle cure that is years away.) This time it is not a new prosthetic leg, but a BBC article titled 'Paralysed man moves in mind-reading exoskeleton'. My friend writes: *Could they do this for your leg? … Hope all well … Catch up soon* in the subject line.

It is an amazing story: a Frenchman called Thibault, who was paralysed in a 15-metre nightclub fall that left him with almost no functional movement below his neck, has moved all four of his paralysed limbs in an exoskeleton, using a brain–computer interface (BCI). Two 64-electrode arrays have been surgically implanted against each side of his motor cortex.

As he imagines walking or moving a hand, the brain activity detected by the implants is fed into a computer, decoded by algorithms and moves the robotic effectors, and Thibault is able to walk across the lab in a 65-kg exoskeleton, controlled only by his thoughts.

There's a video embedded in the article: the white bulky exoskeleton's actuators and motors whir and wheeze as it makes gradual, creeping progress across the white-tiled lab. Halfway through the video there's a still of the electrode, and a plastic model of the brain showing the extent of the two 5-centimetre holes that Thibault has had to have drilled out of his skull. Then we see him heroic at the end of the film, attached to a ceiling harness for stability as the camera pans around him. As the article puts it: 'His movements, particularly walking, are far from perfect and the robo-suit is being used only in the lab.' Then there's a quote from Thibault himself: 'It was like [being the] first man on the Moon. I didn't walk for two years. I forgot what it is to stand, I forgot I was taller than a lot of people in the room.'

Out of interest I go digging for the source. It's a *Lancet* article titled 'An exoskeleton controlled by an epidural wireless brain–machine interface in a tetraplegic patient: a proof-of-concept demonstration'. The article says the aim was to be the first to 'provide tetraplegic patients with an original neuroprosthesis, driven by the patient's mind, in an unsupervised manner, and fulfilling the requirements of chronically implanted devices (wireless, fully implanted, bio-compatible and long term)'. The writing has that particular, privileged style of scientific-research articles, which is hard to follow and a little bewitching, and I have to concentrate hard.

Behind the technical language are hints of what they had achieved, and what it must have been like to be Thibault, the guinea pig at the centre of the study. There are dense pages of results on his performance: percentages and tables of the number of times he successfully achieved a task on the training programmes (make the avatar walk, move the dot to the target, twist the exoskeleton's wrist). And despite the blandness of the writing, I have a picture of Thibault sitting in the lab in front of a bank of computers, sweating with concentration and fatigue as he tried to make things move across a screen using only the power of his mind, imagining again and again, among all the failures, and the small iterative changes the researchers sitting around him had to make to the coding and algorithms so the exoskeleton would match the electrical activity of his cortex; and the joy when Thibault did make it work – the whoops from the team: all over two years, and with a final success rate of about 70 per cent.

There's a reference in the paper to an implant remaining stable for eight months in a sheep's head – the implication being that an electrode of this kind had been successfully and permanently implanted over the motor cortex before, but only in an animal. The stakes were high, and it felt to me like Thibault really was taking a small step for humankind. Early on in the paper there's a brief mention of a *Patient 1* (Thibault is *Patient 2*). Whoever *Patient 1* was, they had two electrodes implanted in the skull, but the device had stopped communicating and so, as the paper puts it with brutal clarity, they were 'explanted and Patient 1 was excluded from the study. The technical problem was identified, and corrected before Patient 2 implantation ...' (no less competitive than astronaut school, then). If Thibault felt like he'd been to the Moon, then he'd done it in the knowledge that *Patient 1* had not made it, and that must have taken guts.

At the end of the article are thirty-odd references. Many of the recent BCI successes are there: patients with locked-in syndrome or tetraplegia controlling robotic arms or a cursor's movement with a variety of brain-control interface technologies during the last decade or so. There are some titles that make sense to me – 'Neural control of cursor trajectory and click by a human with tetraplegia 1,000 days after implant of an intracortical microelectrode array'; 'Reach and grasp by people with tetraplegia using a neurally controlled robotic arm'; 'Fully implanted brain–computer interface in a locked-in patient with ALS' – but others, which I guess focus on the control algorithms, are incomprehensible: 'Recursive exponentially weighted n-way partial least squares regression with recursive-validation of hyper-parameters in brain–computer interface applications'. It's a small window on the vast interdisciplinary nature of so many modern advances.

During the next twenty-four hours I notice the same story popping up in news feeds. It makes it to the top of the 'most-read' rankings. People are engaging with it. Presumably, to interpret and liven up the dense science of the article, the publicity team from the University of Grenoble sent out a press release – important for public-engagement points, and fuel for future funding rounds – and it's landmark research that's worth shouting about. It's spawned at least 130 different articles across the world. There's been some editorial creativity in the range of headlines, dependent, it seems, on what makes for the most enticing clickbait: from the fairly accurate 'Tetraplegic man walks with the help of brain-powered robotic suit', to the less so: 'A brain-controlled exoskeleton has let a paralyzed man stroll within the lab'; and the slightly flippant: 'Tetraplegic patient can now move his four limbs with the help of a badass neuroprosthetic suit'. These sensationalise the advance, in pursuit of clicks, but whatever the hype, Thibault and the team really have made progress.

Sixty-five years ago the first direct electrical stimulation of the human hearing system was performed by André Djourno and Charles Eyriès in France. They implanted an electrode-coil device against the auditory nerve of a man who had lost his hearing. He said he could hear sounds like a cricket or a squeaky wheel. Over the next weeks the patient trained himself to distinguish a few different sounds as short words and said he liked hearing his family again; then, after a month, the coil failed. This experiment is credited by many as being the first breakthrough step towards the cochlear implant.

Deafness and severe hearing loss are most commonly caused by damage to the sensory hair cells in the spiral-shaped cochlea (Latin for 'snail') of the inner ear. When sound enters the healthy ear it causes the eardrum to vibrate, is then amplified by the tiny bones of the middle ear and passes on, through the oval window, to move the fluid that fills the cochlea. As the sound-sensitive hair cells bend in the moving fluid, they create an electric charge that excites our auditory nerve and is interpreted as sound in the auditory cortex of our brain.

A cochlear implant bypasses the missing or malfunctioning hair cells by directly stimulating the auditory nerve with an electrode that has been surgically inserted into the cochlea. Most devices now consist of external parts worn around the ear: the microphone and the processor, and the internal receiver, which sits under the skin behind the ear. The signals pass down a thin wire to the electrode in the cochlea. This is perceived as sound in the auditory cortex. It is a prosthetic substitute and is different from biological sound – it takes time for the brain to learn to interpret this input meaningfully. (A quick google of 'What does a cochlear implant sound like?' gives you some idea of the fizzing distortions of prosthetic sound, but also how much clarity a purely digital signal can create, particularly by some of the modern multichannel devices.)

We've been fiddling about with the effect of electricity on the human body for centuries. In the 1790s the inventor of the electric battery, Alessandro Volta, was experimenting with his new discovery. He decided to put each end of one of his 50-volt batteries into his ears. He wrote it up: 'at the moment when the circuit was completed, I received a shock in the head, and some moments after I began to hear a sound, or rather noise in the ears, which I cannot well define: it was a kind of crackling with shocks, as if some paste or tenacious matter had been boiling … The disagreeable sensation, which I believe might be dangerous because of the shock in the brain, prevented me from repeating this experiment.'

Who the *Patient 1* or *2* was – the Thibault – of the cochlear-implant, is hard to pin down (there were also Russian experiments in the 1930s, which don't get much of a mention in the records). Scientific discovery is slippery and often disputed, but Djourno and Eyriès's 1957 work, which ended when they fell out over ethical and personal differences (should they make their research freely available or patent it?), kick-started further research. Their results were published in the French journal *La Presse Médicale* and led to a handful of news articles across the world. One ran in the *Los Angeles Times* and was handed to Dr William F. House by a patient who was visiting his clinic. House was inspired and, with the help of a team including electrical engineer Jack Urban, went on to invent what is considered the first cochlear implant that could be used outside the lab.

The rise of the new technology brought criticism from many in the ENT (ear, nose and throat) field, who felt the device couldn't possibly restore useful hearing, as the electronic stimulation was just too crude in comparison to natural hearing. Alongside the theoretical disapproval was strong ethical condemnation. The Deaf* community was suspicious and resistant: the cochlear implant would destroy Deaf

---

* 'Deaf' when capitalised is used by those who are part of, and actively engaged with, the Deaf community and identify as culturally Deaf, while 'deaf' with a lower-case 'd' is the medical term, and applies to those with hearing problems, who may or may not identify as part of the Deaf community. Often D/deaf is used as shorthand to describe both groups.

culture and language. If deaf children didn't learn sign language, they would never truly be part of the highly effective and supportive Deaf community and would regard themselves – and be regarded by others – as disabled. By the 1990s opposition to the cochlear implant was widespread in the US, particularly towards implanting the device in children too young to give informed consent. It was an emotional issue, and some in the Deaf community saw themselves as an ethnic group: to them the implant represented 'ethnocide'.

In 1976, when the cochlear implant was still highly controversial, a study showed that all thirteen patients in the US who'd had the implant were significantly better at speech and lip-reading when the device was turned on, and reported a better quality of life. The technology was improving fast. While the experimental devices of the 1960s and 1970s resulted in crude, if useful, sound, the latest multichannel cochlear implants have electrode arrays that, with modern surgical advances, can be placed against the auditory nerve more precisely, resulting in a greater range of sounds; speech processors are smaller and have more powerful coding; and, combined with better microphones, individual voices can be picked out from a noisy background and music can be appreciated.

There are, of course, improvements still to be made. When a cochlear implant is inserted, any natural hearing the patient had, however weak, is destroyed during surgery – there is no going back – so hybrid implants are being developed that could provide a prosthetic substitute while maintaining any residual hearing. And modern cochlear implants still cause some foreign body reaction, biofilms can form and tissues are damaged, so new biocompatible electrodes are being researched, and robotic surgical techniques developed, that limit insertion trauma. At the cutting edge, new arrays are being trialled that excite the auditory nerves with pulses of infrared radiation or light to increase the longevity of devices; and miniaturisation of components and better battery life promise completely implantable devices in the future. And, over the horizon, stem-cell therapies might replace or prevent the loss of damaged hair cells in the first place, perhaps one day providing a regenerative option to hearing loss.

Whatever lies ahead, cochlear implants have already been a transformative technology. Now more than 600,000 people worldwide have a cochlear implant. In just sixty-five years we have gone from no effective therapies for hearing loss, to rapidly developing a technology that lets once profoundly deaf people talk on their phones, appreciate music and pick out an individual's voice from a noisy crowd. And the results have been particularly good for deaf children who receive an implant at an early age – they find it easier to learn speech and language and to fulfil their potential in education.

Those early cochlear-implant pioneers explored what was possible – no matter how rudimentary, how fleeting the bursts of prosthetic sound, they were willing to try it, despite being wired up to utilitarian equipment and confined to the laboratory. There will have been those in early trials and experiments who never heard anything (the implant failed or had to be taken out); they never made the research papers. These people risked the unknown, giving their bodies to science so that one of the greatest advances in medical history could be developed. We should be hopeful that Thibault is in the early stages of something similar, that he has made the first steps towards a brain-reading *badass* exoskeleton that will be *strolling* down the street in sixty-five years' time.

These intrepid guinea pigs are just as inspiring for me as those who climb, cross and explore uncharted and dangerous geography. I look at the risks they take – and for Thibault, it was letting doctors implant two electrode arrays onto the surface of his brain, one of the last properly functioning parts of himself – and I see courage. No one can be sure what complications these pioneers face in the future: infection, implant removal and perhaps, when the operation is irreversible or damages their bodies, whether they have precluded themselves from future generations of the device.

They have raged against the dying of the light. That first patient who went to Eyriès and asked him to try to restore his hearing with electricity, he was raging. And Thibault, and the paralysed people moving cursors or robotic arms with BCI; and the handful of people worldwide who have bionic eyes and can see a few flickering specks in the dark; and Jack and the first osseointegration patients – all raging in different ways. But just as importantly, so are the accessibility developers at tech companies; and the disability activists who chained themselves to the London buses they couldn't access and raged until the 1995 Disability Discrimination Act was passed; and the Deaf community too, for trying to build its community and change society, rather than letting technology alter their bodies to 'normal'. These people show us tech isn't always the solution – we need to change society's attitudes to make the world accessible for all.

Yes, every now and again I will briefly rage – we all do. I will write a letter to a politician, saying I don't think some part of the health system is working properly; or I'll help a charity raise some funds; or I'll suddenly rage that my sockets aren't comfortable and there must be a better way and I'll stomp into the limb-fitting centre. But mostly I don't want to. I want to lead a normal, quiet life that isn't about my disability, and I'm thankful to all those who have dedicated themselves to making medical technology better for the rest of us – those who explore, for Jamie and Hayden and me, so we don't have to. We should be thankful to Thibault. He's never left the lab with his exoskeleton, and it may take many years for others to follow him, but he has helped us step into uncharted territory.

Made in Our Image

While I was in Germany I attended a symposium on assistive and wearable robotics. I'd emailed ahead to see if I could join. The enthusiastic back-and-forth I had with the organiser didn't mention the cost of a ticket, and when I turn up at the table of laid-out name tags, I'm asked which university I am with. I'm just an interested member of the public, I say. I'm handed the card reader and told it will be €300. Too embarrassed to retreat, I pay up and find a seat where I'll be invisible – nearly – but not quite at the back of the room. The thirty-odd people attending are chatting away and for a dreadful moment I think the whole conference will be in German, but as soon as the facilitator stands and rubs his hands together, this pan-European group switches into English.

There are a few presentations during which I have literally no idea what they are talking about, it might as well have been in German (a very maths-heavy talk on control approaches, one on predictive models for human balance control, and another on online planning and control of ball-throwing by a VR humanoid robot …). I notice there is a tendency for the presenters to say the word 'basically' before explaining something that is anything but basic. Every time I hear *basically*, I feel a little further out of my depth and a little more stupid. But mostly the talks are fascinating.

I wasn't sure exactly what an assistive and wearable robotics conference would cover. The most obvious example is the exoskeleton, and many of the lectures focus on the subject, but the field is far wider and seems to include any hardware that repairs or augments the body. And throughout the day we're reminded that this isn't just about healthcare. Slides are flashed up of exoskeletons designed for industry or the emergency services: one of a firefighter carrying a huge coiled hose up a stairwell, assisted by a compressed air-powered exoskeleton; two elderly Japanese farm labourers lifting apple crates while wearing pressurised air-powered suits (the point being made by the speaker that exoskeletons could prevent injury in manual labour and might keep an ageing workforce useful); and a Lockheed Martin lower-limb suit designed for soldiers and first responders – letting them carry more, further, with less fatigue.

My experience as an infantry soldier told me this would be helpful, theoretically – I'd had to carry vast amounts of armour, munitions and comms equipment across dusty places and would have loved anything that assisted – but in practice being dependent on yet another technology and its logistical burden (think batteries, charging and maintenance), not to mention the risk it might fail at the critical moment you wanted to scramble into cover, leaving you exposed under enemy fire like an upside-down insect, made me reckon there was a fair way to go before combat troops would be stomping around in anything like this.

These limitations seem to be why exoskeletons remain tools for rehabilitation rather than devices that significantly help in everyday living. Yes, there are stories of people completing marathons while in exoskeletons, and a handful of pioneers use them day to day, but these are still the exception. Many of the presentations focus on those very specific and technical problems that need to be overcome so the next generation of exoskeletons might let a paralysed person routinely walk out of the house to do a bit of shopping. The hurdles are ones of battery weight and power – we're told how efficient the human body is, performing all day long on a few hundred watts and with only one charge (breakfast); *it really is a miracle*, the scientist says, *we are so far from being able to replicate anything like this* – and, as ever, the challenge of interfacing the tech with the person.

How do you get an exoskeleton to operate in complete synergy with the body? There's the mechanical problem: fitting the suit to each human user safely and comfortably, every one of which will be a different shape and size. Adjustable straps and extensions and flexible materials help, but the joints of an exoskeleton need to mimic the body's vast freedom of movement and various speeds. Take the human knee, for example: it is actually one of the most complex joints in the body. It doesn't simply bend on a single pivot, it's polycentric; as the femur slides over the tibia, the centre of the joint migrates as it bends. An exoskeleton needs to accommodate these sorts of intricacies, supporting us while not putting the biological joint under stress that might risk injury.

But the more difficult problem to solve is the control interface. Short of wiring up the brain (as Thibault had), how do the motors driving each robotic leg forward know what the user wants it to do, especially if the person is paralysed and has no way of initiating control or receiving sensory feedback? And how do you make the exoskeleton balance, stop it tripping over, and adjust to the infinite complexities of the real world?* The small instinctive reactions a human makes to stay upright and navigate our world are fiendishly difficult to replicate.

Moravec's paradox was the surprising discovery by robotics and AI researchers in the 1980s that it isn't high-level reasoning that is hard to replicate; what takes huge amounts of computation is actually the low-level sensorimotor skills we take for granted. We constantly use sensory information and turn it into useful movements; those movements then

---

*In the rehab centre I always felt how well I was doing; the floors were glossy flat, there were lifts and ramps and handles to make life easy. It was only when I went out into the real world, where every surface was uneven and sloped and ever-changing, that I realised how much more difficult mobility would be, and how much less effective my prosthetics were.

influence how we sense future stimuli, in constant feedback loops, which enables us to correct continually for errors and changes in the environment. Preventing a stumble is the sort of instinctive response we've become unbelievably good at – it has been encoded in us over billions of years. However, as Hans Moravec put it, 'Abstract thought … is a new trick, perhaps less than a hundred thousand years old. We have not yet mastered it. It is not all that intrinsically difficult; it just seems so when we do it.'

Take AlphaGo, for instance (the AI program that beat one of the top human players at the abstract-strategy board game Go in 2016, made by the now Google-owned company Deep-Mind). It's an algorithm that decides which moves to make, based on knowledge it accumulated through reinforcement learning. It does this using artificial neural networks trained to identify the winning percentage of each move, which then teach the program to make ever better choices until it is human-beating. All done by playing itself, without any historical data or human intervention. But AlphaGo wasn't a robot sitting opposite a human at the 19x19 Go board, it was an algorithm hosted on vast servers. A human had to move the pieces for it. Because we find moving the pieces so easy (to the point of it being automatic), it seems odd that it is more difficult for us to make a robot that can learn to reach out and pick up the Go pieces than it is for us to make one that can master the abstract problem-solving of a game we have to think so deliberately about.

How to solve these sorts of sensorimotor challenges are the most complicated lectures: endless slides of control algorithms, and gait-pattern analysis, and biological modelling. I'm left with a sense of how the engineering, computing and programming are iterative. Everyone seems to be working on something very specific and complex that, in isolation, is hard to grasp but, when understood in relation to all the other research, starts to make some sense – and this is just one small room of one conference.

After coffee there's a talk from the company that makes my microprocessor knee. I hadn't considered my prosthetic as a wearable robotic, but I suppose it is. Then two new upper-limb myoelectric hand prosthetics are presented.

Next, we're on to robots proper. The word 'perturbation' keeps being used. It's another maths-heavy talk, until we're shown videos of a humanoid robot running on a treadmill: the noise of whooshing actuators and the square feet pounding the rubber. I'm surprised when a researcher's hand comes into frame and gives the robot a push. *So that's what he means by a perturbation.* The robot adjusts by stepping a leg out sideways to counter the thrust. It's shoved again, harder, and the leg flicks further out; it nearly overbalances, but keeps on running. It's an amazingly human-like response.

After the last lecture of the day I find myself unexpectedly in a minibus with a few others, being driven north. I'd been asked – I think by a PhD student – if I wanted to see their robots. I'd said sure, presuming they would be in the next room. It's a half-hour drive to the Karlsruhe Institute of Technology (KIT), one of Germany's top engineering and scientific-research universities. We're shown two robots. The first is AMAR IIIb. He or she (we're not told) has a humanoid upper body of blue panels covering a grey armature of pistons and wired workings. It's twelve years old, but I feel like I'm standing in front of an antique – it reminds me so much of the robot in the 1986 film *Short Circuit*, one of my childhood favourites, that I feel they must be paying homage. AMAR IIIb wheels around a mocked-up kitchen with a chef's hat on, opens the fridge, picks out the right ingredients and makes an omelette. It has slightly absurd white googly eyes, but we're told they house cameras for peripheral and foveated vision and the system is now used for a number of different industrial applications across Europe.

I'd never really understood why anyone would bother creating humanoid robots. I thought it was a deluded pursuit of made-in-our-image entertainment and novelty fembots – surely there were easier ways to reach useful solutions than trying to mimic the strange, spindly instability of human bipedal anatomy. Take a robochef food-processor-type machine that you chuck ingredients in and it chops and cooks for you. It sits on the countertop like any other cooking aid and seems a more efficient and cheaper way of achieving the task. But watching AMAR IIIb wheel over, open the fridge, see the milk, name it, pick it out of the fridge, add it to the eggs it has just cracked and start whisking (even though there is a little spillage, and something slightly laboured about its movements) – added to the lectures earlier that day (the robot on the treadmill recovering its balance, for instance) – makes me realise why all these academics are so interested in humanoid robots. It is another way to learn about the body, and how we can best assist it when it is disabled.

Gait-pattern analysis helps us understand the way we balance, and an anti-stumble algorithm perfected in a humanoid robot might make a lower-limb prosthetic or exoskeleton more sure-footed. And if a robot can interact in our environment fully, it will be more useful – a robochef can cook only a few stew-like dishes; robots like AMAR IIIb can make an omelette, wipe clean the surfaces afterwards and fill the dishwasher. Watching this robot reach into the fridge makes me realise that the technologies it uses to pick up the milk are the same ones used in an upper-limb prosthetic that KIT presented during the morning.

The problem with the most advanced bionic-hand prosthetics (often called myoelectric prosthetics) is the way they are controlled by the human user. As the amputee tenses very specific muscles in their stump, carefully positioned electrodes in the prosthetic socket detect the electrical signals our muscles emit, and this activates the motors that open the fingers or twist the wrist into the right shape to perform a daily task. The drawback with this type of control interface is that it often requires a lot of training and is frustratingly laboured in practice. (It was what was so striking about the super-users racing down the Cybathlon course at the REHAB fair.) The electrodes also need to be in precisely the right spot, the skin needs shaving, and any sweating or movement can reduce effectiveness. Although they look amazing – the stuff of science fiction – the cognitive burden of controlling these expensive prosthetic upper limbs means people often end up not bothering, and around 50 per cent abandon them altogether.

To overcome this, the new KIT hand has a processor, distance sensor and camera embedded in the palm, which can identify a number of objects. Instead of the user having to shape the hand into the right grip with a series of tiring and counter-intuitive muscle movements, the hand could identify the object when it 'sees' it (a banana, a Coke can, a phone, keys, and so on) and naturally moves its fingers and wrist into the right shape to pick the object up, just like AMAR IIIb reaching into the fridge. And because the movements are pre-programmed, the user doesn't have to think about it, and the whole thing looks more natural than even the best myoelectric prosthetic-hand user could manage.

I nearly say thank you to AMAR IIIb when we are led out of its kitchen and down the corridor to meet AMAR-6. Like its older sibling, AMAR-6 is on wheels, with a humanoid upper body. A shiny green shell gives it a more up-to-date look. It's designed for the industrial setting and shows us a bit of cleaning, squirting a disinfectant bottle and wiping a surface. Then it rolls over to a dummy section of warehouse shelving and helps one of the PhD students unscrew and lift out a section. The student has hold of one end, AMAR-6 the other, and while the student walks around, explaining what's going on, the robot mimics his every move. We are invited to take turns raising and lowering AMAR-6's arms, which are 'slaved' to our movements. It's effortless, even though the robot is holding a heavy section of shelf. It's another example of a technology that would be invaluable in an exoskeleton or device to assist the disabled – or any of us as we age, for that matter.

At the end AMAR-6 asks us if we'd like a photograph, and everyone is smiling and taking turns. I ask one of the researchers how she feels about the robot: would she run in to save it, if there was a fire? She laughs. 'Depends how big the fire,' she says. 'I'd feel sad for it if it burnt – but just in the same way I feel sad for anything I like.'

It's my turn and I stand beside AMAR-6, its arm around my shoulder, and take a selfie.

The next morning the symposium continues. There's a new robotic skin, designed to let robots work in closer coopera-tion with humans – an exoskeleton lined with these sensory cells could 'feel' and adjust the pressure it is putting on the body and reduce rubbing, or a robot caregiver could more tenderly lift a patient out of bed. And reverse-engineering some future generation of this tech could perhaps restore spinal-cord-injured patients' sense of touch.

Then two lectures on soft robotics. These are assistive devices with components made from compliant materials, which can be integrated into textiles and clothes. There's a *soft wearable actuator grip*: a glove with rows of bladder cells sewn into it. When the cells are inflated, they push against each other, curling up into a grip – it's very organic-looking, like a fern frond furling and unfurling. A video shows the sort of assistance it can give: the empty glove holds a 9-kg weight. Then we're told about a soft exoskeleton, made of textiles with a flexible tendon that runs up the leg through joints at the ankle, knee and hip. It assists by giving extra power in those first degrees of standing up, when gravity needs to be overcome, or in the stance phase of walking. I think of my elderly grandparents 'oofing' as they struggled out of chairs.

While a rigid exoskeleton can provide more power and hold its shape when assisting the human body – essential in the case of a spinal-injury patient – it is heavy, bulky and needs lots of power. Soft robotics can't give as much support, but are lighter and designed to fit into our daily lives, to be used with a wheelchair, or in a car or armchair. I can see how they might have wider benefit in society, integrated into our clothes to make us stronger or faster, to keep us from 'oofing' as we age.

But perhaps the technology that I'm most impressed by, from the entire two-day symposium, is the simplest. It has the graveyard slot, when PowerPoint fatigue has set in and there's a certain restlessness in the room – and I suspect it's not as interesting for many here who are devoted to the high-tech. It's a prosthetic hand that grips by simply being pushed against an object; each finger adapts to the shape of the object, so it can pinch or grasp and can pick almost anything up, but it's purely mechanical – no heavy motors and batteries, or clever control algorithms. And because the user is pushing against something physical, the user feels more feedback. It's one of the lightest prosthetic hands in the world, is cheap to produce and could be used in low- and middle-income countries, where (let's not forget) the majority of amputees live. The young professor from Delft University of Technology who is presenting describes it as a smart technology. 'You wouldn't use a Formula One car to do the grocery shop,' he says. It's a good-natured jibe at some of the ideas that have been presented and gets a laugh.

I've just spent two days listening to very complex, expensive solutions, and the one that makes most sense to me is a smart-tech solution – it is still a cutting-edge mechanism, inspired by the human hand, but there are no batteries, it is 3D-printable and intuitive to use.

And then the symposium is over and I am heading for the airport.

A few years back I met Aldo Faisal, a professor of AI and neuroscience, in his office at Imperial College London – whiteboards covered in equations and algorithms and feedback loops, quite sparse, a little messy, with a few incongruous artefacts dotted about: a book about a minimalist artist, and another about a fashion designer, which I liked to think might give him tangential inspiration. I'd asked to meet him after I attended a lecture he'd given on a new Artificial Intelligence Clinician he and his team were developing to help treat sepsis (a leading cause of death worldwide, and most common cause in hospitals). This AI would sit alongside a patient in intensive care (no robots here, just software running on a pretty standard monitor, of the type you might already see) and measure vital-sign outputs and therapeutic inputs and recommend the best treatments. Professor Faisal had walked about the front of the lecture theatre holding the laser pointer, his enthusiasm bringing us along with him, and it was probably as close as you'll get to the young rock star professor of the movies. I was a little in awe.

I'd gone along because some of my experiences in hospital had been grim. For the most part the nurses and doctors had been kind and brilliant – but human error, fatigue and maybe, once or twice, inexperience had meant that I'd often (when I was conscious anyway) asked about a drug I was being given, and whether it was the right dose or right time. The medics had frowned – 'I'll just go and check.' And once I was given something when I shouldn't have been, and my kidneys hurt and I felt dreadful. It gave new meaning to the saying 'You have to own your own recovery' (which is the best tip I can give anyone going through complex medical care). And then I'd got a fungal infection – no one's fault, just bad luck – and they'd had to remove my other leg.

Could an AI have performed better and spotted, in all the information produced by the beeping monitors I was hooked up to, the signs of infection before the only option left was amputation? In his lecture Professor Faisal outlined how challenging and time-consuming it was to take huge datasets from intensive care units, which had tracked every intervention and observation in thousands of patients, and clean them up ready to be used; how difficult it was to cancel out the noise of human behaviour and biology; and how bad data-gathering often is, if it happens at all – getting the dataset ready from which to create your AI seemed to be half the battle.

But once Professor Faisal's team had it, they created an AI that extracted knowledge from patient data that far exceeded the lifetime experience of a human doctor and, using reinforcement learning, they could predict mortality and suggest the best treatment. While a good doctor can make judgements from about five or six different parameters, Professor Faisal told the lecture room, the AI Clinician could track around twenty different variables in order to make recommendations. It had also uncovered a tendency for doctors to increase the dose of drugs as patients became sicker, which had little impact and even caused harm.

In his office Professor Faisal had eaten his sandwich and, between mouthfuls, explained more and even stood up to wipe away some old diagrams from the whiteboard and scribble a quick perception–action loop. His interest was creating AI by reverse-engineering from first principles the algorithms that drive human behaviour. He said his work was almost science fiction; he let the rest of the team worry about how it could help the end user. Slightly out of my depth, I'd said how brilliant I thought all this AI stuff was. But he'd cautioned me: remember those passenger jets that crashed shortly after take-off; the on-board AI was being given faulty data from sensors that said the aircraft was stalling, when it wasn't. The AI bypassed the pilots and, much as they fought to bring the nose up, the AI kept forcing it down. AI is only as good as the data we give it.

I'd left his office and walked through students queuing for their next seminar and been reminded of a conversation I'd had with one of the developers of my microprocessor knee. He'd told me how responsible he felt when they let the first knees out to the patients. 'In the lifetime of your leg it has billions of decisions to make,' he'd said. 'It takes information from all its sensors, adds some historic data and feeds it

147

into our magic control algorithm, which is a cascade of very elaborate control approaches, and decides what to do next – take a small step, a big step, stop a stumble, yield down some stairs, et cetera.' Then he'd asked me: 'If one out of a million decisions is wrong, do you think it will be a problem?'

I said I wouldn't have thought so.

'Let's say it's two hundred billion step decisions, for all our new knees out there in the world being used by amputees over a year. If one out of a million decisions is wrong, that is two hundred thousand missed steps; if one out of ten missed steps leads to a fall, that would be twenty thousand falls; and if one out of ten falls leads to an injury, that's two thousand injuries. If one in two hundred injures is fatal, that's … well … not good. The health of the end user is a great responsibility for the control engineers. We have to get the whole system right. There's no margin for error.'

We've long wanted to play God. It's thought the ancient Greeks created automata that mimicked animals and people; da Vinci made plans for an artificial man – a medieval knight powered by cranks and pulleys and cables; the French inventor Jacques de Vaucanson created a mechanical duck that wowed crowds in the 1730s, flapping its wings, eating and defecating; and Thomas Edison brought a talking doll to market that was so uncanny it probably started the scary-doll horror genre. All these sit somewhere between magical illusion and mechanical wonder.

Now we can create truly useful robots, which are able to interact with us and our environment. In medical settings they can help treat patients with contagious diseases, clean wards and assist us with surgery. In nursing and retirement homes, experiments have shown patients developing an emotional attachment to humanoid robots, helping with dementia and Alzheimer's. The toy company Hasbro sells a lifelike robotic companion cat that can purr, roll over and blink when stroked. In the future, robots will take on more of these roles – assisting hospital staff and becoming care-givers to the world's ageing population, lifting, cleaning, interacting.

Robots inspired by human biology can function in a world designed for us, but they also become the prosthetics and assistive tech that can repair and replace our damaged bodies. And the human brain (so often touted as *the most complex object in the known universe*) becomes the inspiration for scientists like Professor Faisal who are trying to mimic its capabilities and create general intelligence. AI already helps us make decisions from data that we couldn't hope to crunch, and makes assistive tech more intuitive. Where robotics and AI converge promises new technologies that might integrate with us more seamlessly, lessening the physical impact on our bodies and the cognitive burden. This will be particularly powerful for the hybrid humans of the future.

Extended cognition is the idea that our mental processes are extended into our environment. Andy Clark and David Chalmers first described the idea in the 1998 paper 'The Extended Mind', which opens with: 'Where does the mind stop and the rest of the world begin?' When we undertake a mental task we use the things around us: for arithmetic, for instance, we might count on our fingers; for longer sums, use pen and paper; and get out the calculator for greater speed or complexity. We are constantly using 'the general paraphernalia of language, books, diagrams, and culture'. In using one of these external entities (the pen and paper, say) we create a *coupled system* in which we delegate part of the task to the technology. This coupling counts as a cognitive process, even though it's not wholly in the head. If you remove the technology (take the pen and paper away), our ability drops, just as it would if you removed a part of our brain.

The paper uses a thought experiment: Inga wants to go to an exhibition at the Museum of Modern Art, New York. She remembers that the museum is on 53rd Street, walks there and goes in. The belief that the museum was on 53rd Street was in Inga's memory, ready to be accessed. Otto also hears about the exhibition and wants to go. But Otto suffers from Alzheimer's. To overcome his disease, he carries around a notebook in which he writes any new information he learns – it stands in for his biological memory. Otto refers to his notebook and it tells him the museum is on 53rd Street, so he walks there and goes in. Was Inga's and Otto's belief that the museum is on 53rd Street any different just because Otto's memory was held external to his mind, delegated to his notebook? Perhaps not – it had the same result. If you replace the late 1990s references in the paper – the *calculator* and *Filofax* – and make my smartphone and prosthetics some of the cognitive resources I *bring to bear on the everyday world*, then I am very much a *coupled system*.

I was once asked by a friend what it was like when my leg broke and I couldn't get a replacement. In reply, I asked how she'd feel if she lost her phone. (She had all the photos of her children on it, all her work contacts and emails, all her notes and passwords, and wasn't sure it was backed up.) She said, 'I'd be distraught … devastated, and I don't know how I'd get anything done.' She also talked of the anxiety of not being in instant contact with people – that was scary. 'It would be like a part of me was missing,' she said. 'That's because your phone is like a prosthetic for you,' I said. Being suddenly without my legs creates the same feelings for me. The only difference is that sometimes, when we lose our phones, after a day or two we realise that we're going to be okay and it's rather nice not being coupled to it any more. I don't get that epiphany when my legs aren't working.

This coupling is intensifying as the technologies we use get more sophisticated. The advances in AI and robotics will make for the most intense couplings. As with all technology, there will be benefits and costs: some we can see coming, others are over the horizon. But the disabled will be at the forefront of testing what the future might be like – and where the human ends, and assistive technology begins, opens up all sorts of legal and ethical questions: what if you snatched Otto's notebook away from him and tore it up – would that be the same as damaging Inga's brain?

This isn't a problem for science fiction; it is already here. In 2009 a 6-foot 6-inch, sixty-three-year-old tetraplegic Vietnam veteran, who had almost no functional movement of his legs or arms, had his mobility assistance device (an electric wheelchair he was completely reliant on) broken by an airline on a flight from Miami to Puerto Rico.* He didn't receive a replacement for a year. Bedridden and without his assistive technology, he had to hire in people to help him. He claimed against the airline for the extra costs. But the airline wouldn't pay, saying that because it was a baggage-claim incident, he wasn't entitled to any compensation. They hadn't damaged the man, they said, only his device – likening it to a car accident where the owner was not in the car. But the man managed to prove legally that the assistive wheelchair was his prosthetic, functioning as an extension of his body, and that by harming the device the airline had harmed him, and he won the case. As we become more enmeshed with technology, it's increasingly difficult to separate the person from the devices they rely on – who would Stephen Hawking have been without his mobility and communication devices?

---

*The lawyer who represented him wrote up the case in a paper titled 'Case Study: Ethical and Legal Issues in Human Machine Mergers (Or the Cyborgs Cometh)'.

The Cyborgs Are Coming

It is one of those days when the rain sets in and the wipers are a ker-kushing metronome, and you might as well be ploughing through a storm at sea as driving on the motorway. Andy is telling me about an experiment he is working on, but pauses as he lines up the car and then sends us through the spray billowing from a truck's undercarriage, and we're out the other side among the pairs of red brake lights.

What would Alice look like to Bob if she was falling into a black hole and Bob hadn't entered it yet? It's a thought experiment, which I can only see in images. Alice is a little girl stretching across the event horizon and Bob, who I've imagined as her brother, is screaming for her to come back. But Andy, I suspect – by the way he is trying to explain it to me – is seeing it only as an equation (Alice and Bob are just a and b to him, connected by symbols and calculations). It's about the stretching of light and what colour-shift would happen, or something. I know maths is a language of the imagination, and I am nodding and making encouraging noises, but I don't think I will ever be able to fully comprehend the complexity hidden in the terms he is using.

It's a few hours' drive and, to pass the time, I have opened Andy's university profile page on my phone and I'm going down the list, asking him about his research interests. We're in his Peugeot coupé. Soon after we had set off from his university campus I had pointed at the digital display on the dash, reading thirty-five degrees, and said, 'That cannot be right – it's freezing.' He'd said (a little hurt) it was broken. It's an old Peugeot and he liked it – the way it felt; even though it's hard to get replacement parts, he didn't want a newer car. He'd gestured to a few more things around the cabin that were broken – mostly electronics. It had been pretty cutting-edge when it was first released.

He pauses again. He's lining up to pass another lorry; I feel like we're in a slightly rattly, late 1990s starship and tighten my hold on the door grip – *punch it!* – and Andy is sending us into the swirling nebula.

We're in search of the first cyborg.

How I came to be in a car with a theoretical physicist travelling down the M4 has to do with a boy in the street, probably ten, who asked me, 'Are you a cyborg, or something?' I smiled and said what I normally say: 'No, half-robot.' He was with his friends after school and they laughed at me, but mostly at the boy for asking a stranger a question. I wasn't actually sure if I was a cyborg, so I looked it up later.

The word 'cyborg' is a blending of *cybernetic* and *organism* and was coined during the space race. The scientist Manfred Clynes first used it in the article 'Cyborgs and Space' for the September 1960 edition of *Astronautics* journal. He was thinking about ways we might explore extraterrestrial worlds. For him, we shouldn't try to make habitable environments up in space (spacecraft and space stations), but rather should adapt ourselves to survive the inhospitable vacuum by amalgamating the human with technology – then we'd truly be free to explore. A very practical, if speculative, thinking-outside-the-box idea, and all before Yuri Gagarin first made it out there.

But in sixty years the term *cyborg* has taken on many meanings. My search results were littered with art, comics and the movies. It's become a modern myth filled with fear, hope and the monstrous. A twenty-first-century parable about freeing ourselves from our frail bodies; the chance of immortality set against the stalking menace of technological enslavement. A modern Midas, with a twist of the werewolf. I had sudden images of our ancestors trying to entwine themselves with nature for empowerment – shamans dancing about the fire dressed in deerskins and antlers and communing with another spiritual dimension. The cyborg is ripe with the same sense of mystery.

Early among the search results was the Cyborg Foundation. It's one of those very visual sites, where a montage of slick graphics and videos is collaged together and embedded under the text (cells divide, we fly through the inside of a microchip, strands of DNA unravel, fading into African wildlife, and then a space station and the Earth revolving). It's great to look at. And positioned up front is: '*OUR MISSION IS TO HELP PEOPLE BECOME CYBORGS, PROMOTE CYBORG ART AND DEFEND CYBORG RIGHTS*'. Scroll down and there's a *DESIGN YOUSELF* section with a step-by-step wire diagram, then further down a *Cyborg Bill of Rights V1.0*, which includes: *freedom from disassembly*, *equality for mutants* and the *right to bodily sovereignty*. It's as much artwork as website.

It was set up by two cyborg artists, Moon Ribas and Neil Harbisson. Harbisson has an antenna osseointegrated into the back of his skull, which bends forward over the top of his hair to dangle in front of his forehead. (It reminds me of an anglerfish lure.) He has achromatopsia – he's only ever seen in black and white. At the end of the antenna is a fibre-optic tip that detects colour; the implant then converts the colour's frequency into vibrations that he feels and hears in his skull. So if an orange is held up to the tip, Harbisson senses the colour by the pitch of the vibration.

He developed this device – the 'eyeborg' – while at university and, after a few refusals on ethical grounds, found a surgeon willing to implant it. Soon after he'd implanted the eyeborg, Harbisson's passport came up for renewal, but the UK Passport Office rejected the photo he'd sent, with the prosthetic dangling over his head. Eventually, with help from doctors and friends, he persuaded the Passport Office that he identified as a cyborg and that the implant should be counted as one of his organs. The photo was accepted, and the press declared him the first cyborg to be officially recognised by a government. There were quite a few articles about Harbisson;

whether or not he's the first cyborg, he certainly seems to be a cyborg celebrity.

Moon Ribas is a childhood friend of Harbisson. To become a cyborg, she had a device implanted into both of her feet that connects her to online seismographs. She can perceive the seismic activity of the Earth as it happens – those little ripples and quakes the rest of us are oblivious to. It's a completely new sense, she's said, one that doesn't make her feel closer to robots or machines, but to nature: the more she can feel the Earth moving, the more empathy she has.*

By some of the definitions I found in the results, I do seem to be cyborg: *a being with both organic and biomechatronic body parts … a creature that is part human and part machine.* And I'd met disabled people over the years who had proudly introduced themselves as cyborg, claiming the word as part of their identity. A quick look through academic articles shows how the word has been used for meaning and metaphor in all sorts of fields – anthropology, identity politics, ethics, sociology, architecture. There are papers that describe people who have medical devices such as implantable cardioverter-defibrillators or cochlear implants as 'everyday cyborgs'. And then some articles argue that humans are so enmeshed with

---

*And in one of their more recent collaborative projects, Harbisson and Ribas have connected to each other. They both have what they call a *transdental communication* system: a tooth implanted in each of their mouths. When one of them presses a button, they can send a signal in Morse code to the other cyborg's tooth, which vibrates and communicates the message.

technology we're all cyborgs already. It's such a potent symbol that anything seems to go. 'Sure, I'm a cyborg,' I could have told the cocky boy in the street.

But other definitions, which focus on enhancement of the human (*a person whose physical abilities are extended beyond normal human limitations by mechanical elements built into the body*), seem to suggest that I don't quite fit into the category in the way that Ribas and Harbisson do.

Later that week I mentioned all this to my father-in-law, who said, 'I know a friend of the first cyborg. You should meet him.'

'Harbisson?'

'No, Kevin Warwick.'

I navigate Andy to the suburban tree-lined street, and he turns us into the driveway of a 1930s semi-detached house dripping with rain. Somehow I didn't expect to find the first cyborg living on a residential road just outside Reading. I'm not sure what I'd expected: that maybe you'd be able to distinguish a cyborg's residence from all the others on the street. But it's all very normal and rooted in the world of the familiar – the past doesn't disappear when the future arrives. Even so, some of that mythology is acting on me and I feel a little intimidated; popular culture and science fiction are ramping up my expectations – a cyborg will be difficult to talk to, a little inert or mechanical, even aggressive and dismissive of a slow and yet-to-be-upgraded human.

In fact a tall, thin, sixty-something grey-haired man, in a loose shirt, stoops through the porch. 'Come in,' he says and shakes my hand and leads us into his living room. After Andy and Kevin have caught up, Kevin makes us tea and then sits opposite us in an armchair. My first impression is that he is smiling (and smiles a lot during the hour and a half we spend together). More than that, he is very alive, and his legs cross and re-cross as we talk. Just very human and warm – nothing of the myth. When I mention that the internet pages that claim Kevin Warwick as the world's first cyborg have been sifted down and replaced by ones in which Harbisson gets the honour, he laughs – it doesn't matter to him.

'I did my experiments before Neil was around,' he says. 'He's got it on his passport, though. He's a great guy. We meet up at events every now and again, when we're presenting at the same time. Neil's an interesting case. He challenges different definitions of the cyborg. Having something extra that goes beyond the human norm is the definition I believe is most useful. Neil was colour-blind, so his implant overcame that; it's a sort of therapy for his impairment, but also allows him to perceive infrared and ultraviolet light the rest of us can't. That's interesting.'

And while Harbisson describes himself as a cyborg artist, it's very clear from the way Professor Warwick talks he is a scientist. His first cyborg experiment, Project Cyborg 1.0, was in 1998. He implanted a silicone RFID-chip transponder into his forearm. As he moved through the Department of Cybernetics at Reading University, the signal emitted by the chip was monitored by a computer. Doors opened for him, and lights, heaters and computers switched on automatically. I'd read a bit about it, and I find myself trying to see the scar on his wrist, but it's very dark in the room. Kevin hadn't turned on the lights, and with the curtains half-drawn and the dank rain outside, we're sitting in the half-light. I suppose it's the most cyborg thing about the whole meeting – we don't need to see each other properly to transfer information between us.

There are thought to be around 6,000 people in Sweden with RFID implants. The implants are grain-of-rice-sized and normally inserted into the fleshy triangle between thumb and forefinger. You can make payments in shops and on public transport with a swipe of the hand, and set the implant up for keyless entry to the home or office. Radio-frequency identification (RFID) has been around since the 1970s. It uses Near Field Communication (NFC) technology, so any device that supports this (smartphone or contactless card reader) can communicate or activate an application when the RFID is in close proximity. These small chips are everywhere now: in security passes, credit cards, toll-road tags and passports.

It was reading about Kevin Warwick (and seeing an identification chip injected into a pet) that gave professional piercing artist Jowan Österlund, the CEO of Biohax International, the idea to develop the tech for the general public. Österlund sells an 'install' for around $180. And while Kevin had proved the concept in a respectable scientific setting more than twenty years ago – with all the rigour, ethical sign-off and groundwork you'd expect – RFID chipping's route towards the mainstream has ended up being through the fringe world of piercing, tattoo and body modification.

Body hacking sits as a sort of subculture within a field called *biohacking*.* It's a branch of the health-and-lifestyle market and grabs the attention by playing on the idea that the body can be 'hacked' for better performance. It seems to include almost any new fad: diets such as intermittent fasting or exotic-sounding specialist supplements; nutrigenomics (tailoring nutrition to a person's specific DNA); red-light therapy, in which you shine near-infrared light at the skin to encourage natural metabolic processes; audio-entrainment (playing functional music to relax and restore the brain); cryotherapy (exposing the body to incredibly cold air); and more widespread and accepted practices like meditation and positive psychology. Some of it has a strong evidence-base, some not so much, but it's a powerful marketing tool playing on the 'hacker ethic' – that information should be free and shared, that we can change our lives for the better, and to distrust authority voices and opinions (that might be scientists, government advice or evidence-based research). If you remove the pseudoscience and marketing hype, it's about self-improvement and agency, taking control of your health and increasing the chances of a long, disease-free life. Well intentioned and admirable, when safe and backed up by evidence.

As the extreme outliers within this community, body hackers (sometimes called bio-punks, or grinders) try to improve themselves with cybernetic devices or by altering the body's chemicals and genetics – all experimental, and from

---

*'Biohacking' made the *Oxford English Dictionary* new-words list in 2010.

161

labs in the basement. The goal is to extend human capacities by hacking themselves with affordable off-the-shelf kit – to become cyborgs. Installing RFID chips in your hand is now one of the more tame procedures of the body-hacking community. Groups, collectives and small companies around the world are running do-it-yourself experiments, and the options for modification have expanded. Take your pick (most of it can be bought online): if you want a circle of five LED lights to shine through your skin in time with music, you can get Grindhouse Wetware's rather uncomfortable-looking North-star implanted in the back of your hand; if you want to feel the electromagnetic pull of speakers and hard drives, or magically move paperclips across the table, a biosensing magnet injected into your fingertip from Dangerous Things might be for you; if you want always to know how you are orientated, a North Sense device from Cyborg Nest pierced into the skin over your sternum will vibrate every time you turn through north; and you might even want to get in touch with someone at CYBORGASMICS, whose website says the Lovetron9000 vibrating pelvic implant is coming soon.

And if hardware isn't your thing, there is also a whole group of body hackers engaged in do-it-yourself synthetic biology. Just as the first PCs gave rise to hobbyists coding, hacking and building computers from scratch, so the new hype in genome editing has led to a rise in do-it-yourself genetic engineering. Open-source techniques and increasingly affordable equipment have helped, but it is mainly possible because of CRISPR, a gene-editing technology that can find and cut a sequence of DNA in a cell and replace it with another – a process that used to take an expert laboratory months can now be done in days at home and at a fraction of the cost.

NASA scientist-turned-biohacker Josiah Zayner is one of a few high-profile characters in the field who have tried to genetically engineer their own bodies. He injected himself (between swigs of whisky) on the stage of a synthetic-biology conference with a home-made therapy he said would modify his muscles' genetic expression and make his arms stronger. There's no evidence it worked and it looks more like a marketing stunt.* His company sells a Genetic Engineering Home Lab Kit for $1,440, which includes everything you'd need to start playing around with your own genome (or to make glow-in-the-dark beer). With it you can rewrite the body's biological instructions, if you don't mind the potential risks of fiddling around with your own genetics.

The expansion of the body-hacking scene in the 2010s has had a little of the wind taken out of its sails in the last few years. Conventions have been suspended; new devices and therapies haven't made it to market as quickly as promised, hit by snags in development and funding; and some of the first-generation implants have stopped working, or are faulty

---

*He also tried to replace his microbiome (he suffered from irritable bowel syndrome). He first removed all the bacteria from his skin and gut with antibiotics in a sterile hotel room, then substituted it with a friend's microbiome. That meant rubbing himself with cultures he'd taken from his friend's skin, and ingesting his friend's faeces.

and are being removed. The number of people wanting RFID implants seems to be slowing – the rest of us haven't piled in behind the early adopters. Pandemics and politics may have something to do with it. Regulations have been tightened by governments concerned about safety, fraud and identity theft. And even a general sense of mistrust pervades, stoked by conspiracy theories about mass vaccination programmes being a cover for chipping us all for control and state surveillance.

But it also has to do with the technology. It just isn't useful enough yet. Yes, it's slightly quicker to move through the Swedish transport network; yes, you can't lose an RFID implant; some people have claimed it is eco – fewer credit cards mean less plastic waste – but you can't easily upgrade an implant, can't repair it when it goes wrong, and taking it out is far harder than putting it in. (I've tried to remove shrapnel I thought was just beneath the surface, and quickly wished I'd never started. It was much deeper, and much more painful that I thought it was going to be.)

A smartphone can do everything an RFID chip can, and more. Human enhancement is still easier with wearables. Smart glasses, prosthetics, exoskeletons and health-and-fitness monitors are all solutions that can be upgraded and removed, when needed. Imagine a set of smart soft-robotic clothes – a body suit – that could stimulate your skin in response to an input (warning of a rise in the air pollution on the street you are walking along, let's say); and that had sewn-in artificial muscles to assist with walking and running; and could keep tabs on your heart rate, blood pressure and activity levels, and alert you when something wasn't right or that today you needed to eat more fibre, or should drink another glass of water; that could keep you at a temperature of your choosing, even on a hot underground train; and hardly ever needed charging because it harvested energy from your body movements; and you could take it off at the end of the day and chuck it in the washing machine. Why permanently modify your body if a wearable offers a safer, upgradable and more effective solution than an implant?

'I'm surprised about the RFID,' Kevin says. 'My experiment was over twenty years ago. Back then I thought it would be picked up quickly for uses like a passport. It would be perfect for walking through queues quickly, but it hasn't happened. You can pay for the trains in Sweden, but there aren't many other applications.'

'I think most of us will have an implant some day,' I say. I worry I'm saying this just because I think it's what he wants to hear. I do really believe this – it's not such a moral or emotional leap from a matchstick-sized implant that lives in your arm and delivers a small dose of contraceptive hormone, or a cosmetic implant that increases your breast size, or any of the medical health devices that the 'everyday cyborgs' already have to keep their hearts in check, or monitor insulin levels, or reduce peripheral pain. I do think the technological growing pains will be overcome and we'll be jumping on board, perhaps with a mix of implanted biometric sensors and external wearables. But I suspect I'm more cautious than Kevin about how soon that will be.

There's something of the prophet about Kevin, sitting there in the armchair haloed by the light of the window behind. He showed us the future and is surprised that we haven't followed him there. He re-crosses his legs. 'But the second experiment – that's been used in the US for paralysed people and tested out in different ways. It's even got one woman feeding herself, I believe. But it's all still experimental and in the laboratory. I thought we'd be further on by now. Far more hybrid.'

Project Cyborg 2.0 was a few years after the RFID chip. Kevin had a BrainGate BCI electrode array surgically implanted into the median nerve of his left arm. (While the RFID was passive, this experiment physically wired Kevin's nervous system with a computer.) Using the neural interface, he was able to control an electric wheelchair and make a robotic hand open and close in the lab in Reading. He also flew to Columbia University, New York, to control the same robotic hand over the internet, and received neural stimulation feedback from sensors in the robot's fingertips. The experiment was a forerunner to much of the BCI research that is ongoing today. It's how Thibault controlled his exoskeleton, and has restored movement in the paralysed and translated neural activity into speech, writing and control of a tablet computer – life-changing for those with locked-in syndrome.

In the car on the way, I'd asked Andy about Kevin's reputation among scientists. I'd read a few articles describing Kevin as a *maverick*, and his experiments as *not much more than entertainment*. And I'd seen a few of the pictures from the experiment, close-ups of the surgery, the grey designed gauntlet that held the interface to his wrist and looked as if it was from a TV sci-fi set. Some of the emotive language Kevin used in videos and interviews from the time wasn't what you'd expect from an academic, and I could see how this might provoke others in the field to attack him. Andy felt this was unfair: yes, with the importance of evidence-base and impartiality in modern science, performing experiments on yourself was always going to be controversial. 'But Kevin's done much more than the cyborg experiments, you know.'

As if answering this, Kevin says, 'It's probably true I wanted the publicity – it helped with the funding, and I felt it was important to use words people could understand when explaining the work – but mainly I was doing it for the science. If my going out on a limb makes some people uncomfortable, does it matter, if it increases the understanding that improves the blind's ability to see, or of the paralysed to walk again?'

'I don't think so,' I say.

'It's all had to be meaningful.'

Despite the controversies and 'Captain Cyborg' headlines the press loved at the time, it's a reminder of what sets Kevin apart from the artists, hobbyists and body hackers that he is now so often mentioned alongside. Almost all of his research is about improving the lives of the disabled. From early in his career he was using robotics and computing to create aids for the disabled: a walking frame that gradually reduced its assistance as the patient improved; a system for sending simplified sign language over the landline telephone; and a self-emptying bath that detected epileptic seizures. Even if he was controversial, he was invited to present the Royal Institution Christmas Lecture in 2000, which was titled 'Rise of the Robots', and his biography now includes a long list of awards, honours and honorary doctorates.

There's a lot that is fascinating about the Project Cyborg 2.0 experiment, but I get the impression Kevin wants to move on. (I suppose in the same way I get bored of always being asked about my prosthetics, the cyborg experiments must follow him around, as if that's all he is – it must get a little wearing.) And sure enough, he seems more animated when we talk about what he is working on now.

'AI is like magic for real,' he says. He is explaining his research on Parkinson's disease, using data from deep brain stimulation devices. 'The DBS electrodes are there to push current into the brain and stop the tremors,' he explains, 'but you can also take current out to see what's going on. We use AI to model the brain and predict when the tremors are going to start – to create an early-warning system, that was the goal. But what's interesting is the power of AI to classify. Doctors find it hard to diagnose the type of Parkinson's someone might have. They can only see how the disease presents in the patient – and the shaking and tremors look pretty much the same in everyone. The brain signals that cause those symptoms are actually varied and may be completely different diseases. As a human, we get a limited perspective, we only see what it looks like in the outside world; we found the AI can classify the signals into groups we'd never known about, which then lets the surgeons use more appropriate treatments. As with any neurological problem, you've got human simplification going on because we don't understand, so we might say "This is Parkinson's disease" but, in fact, it could be one of a handful of different things – you get overlap with diseases like dementia. AI can help us diagnose much more accurately.'

Our discussion about his Parkinson's research moves to the future again. 'You can't completely predict the outcome,' Kevin says. 'That's why AI is powerful. But it's also a question of control.' He holds his hands apart as if suspending an invisible cat's cradle. 'The benefits of AI also open up the chance that it will act against us. I do think it could be dangerous.'

Most of us have some sense that AI is a powerful tool. In the medical setting it can predict the onset of tremors in Parkinson's and minimise the insulin dosage of implanted diabetics pumps so they last longer. It's particularly well suited to image-classification tasks, identifying skin cancers from photos, breast cancer from mammograms and the onset of eye disease from retinal images (Professor Faisal's AI Clinician helping treatment in ICU). An AI developed by Google's DeepMind, called AlphaFold (a descendant of AlphaGo), predicted a protein's three-dimensional shape from its amino-acid sequence, helping to solve one of biology's greatest problems, and may well revolutionise many areas of science. It's an incredibly powerful analytical tool for researchers, helping us extract insights from huge amounts of data that humans alone couldn't hope to crunch.

And we also have a sense of AI's limitations. While some

of the current AI doctors are better than your average middle-career doctor, they can't outperform the best senior doctors – the problem being that if future doctors rely too heavily on AI, they might never accrue the experience needed to become expert, and where would that leave us? There's also a creeping uncertainty. The most advanced AI can help us identify the shape of a protein, or drive an autonomous car, but their systems are now so complicated – using deep learning – that the people who designed them struggle to understand how and why they arrived at any single judgement, and that is disconcerting. The next questions seem to be: will AI surpass our intelligence? Will it go rogue and become a threat to us?

It's easy to dismiss the idea of an uncontrollable AI that wipes out humanity as the shrill cry of overreaction – something so far in the future that we shouldn't bother worrying. But it's probably worth listening when some of the best minds are warning us, as Stephen Hawking did shortly before he died, when he made a speech in 2017 at Web Summit, Lisbon: 'AI could be the worst invention in the history of our civilisation,' he said, 'that brings dangers like powerful autonomous weapons or new ways for the few to oppress the many … AI could develop a will of its own, a will that is in conflict with ours and which could destroy us. In short, the rise of powerful AI will be either the best or the worst thing ever to happen to humanity.'

'I agree with people like Stephen Hawking,' Kevin says. 'We have to look to the future. But how AI could enhance a human, I have to admit, is most exciting for me. We showed twenty years ago that you could link the nervous system

– your brain – and control a robotic hand. If we can link the human brain via a neural implant to an AI, it could give us access to entirely new realms of information and experience, and it might also be our best chance of controlling AI.'

I find myself telling Kevin about an argument I'd had with a friend. We were discussing Elon Musk's company Neuralink, which is trying to develop a brain–machine interface: an ultra-thin neural lace with thousands of electrodes that could be implanted into the skull. The goal in the short term is an upgrade to the BCI (like the BrainGate Kevin used), with improved therapeutic benefits for the paralysed and those with neurological disorders – but at the Neuralink launch event, under the bullet *Create a well-aligned future*, Musk also spoke of a device that could achieve a human symbiosis with AI. 'Even in a benign AI scenario, we will be left behind,' he said. 'With a high-bandwidth brain–machine interface, we can go along for the ride.'

I'd been pretty adamant with my friend that there would be too many technical and biological challenges. He'd said: yes, but let's pretend. And so I'd rambled on about humanness and ethics, and how we would want to hold on to the body as sacred, to feel our place in the world. But my friend had argued: what if he bought a neural lace that cost fifty grand for his children and they were funnier, cleverer and healthier than my children, and got into top universities and had better jobs, lived happier lives? Would I not want to implant a 'lace' in my children? My first thought was about the fifty grand, but also about the horror of a technology that would truly upgrade a human – the potential for all sorts of

171

societal inequality and the damage it might cause. So, no different from pretty much all the technologies already in your life, my friend said. He had a point: so many technologies made our lives better, but also created inequality – not least the prosthetics I was wearing to walk alongside him.

'It's when an application becomes irresistible,' Kevin says to this. 'Like laser eye surgery, or the first DBS devices; they were seen as ethically wrong and dangerous thirty years ago. Even the cell phone – everyone was scared of signals going through the body and we were told it wouldn't be practical; but now everyone has a phone and we don't mind about the signals because the technology is so powerful. I think it'll be the same with neural implants.'

Kevin is loosely part of an intellectual movement called transhumanism. It's a philosophy that studies the benefits and dangers of technologies that could overcome human limitation. At the heart of the movement is the possibility that all disability, disease and ageing will be eliminated – in the future we will be freed from the weak, badly designed bodies that evolution has given us. If body hackers are the grassroots foot-soldiers of the movement, unwilling to wait for the future to arrive and trying to upgrade themselves with what we have now, people like Kevin and Elon Musk are the sky marshals showing the way.

The 'Transhumanist Manifesto' by Natasha Vita-More, now on its fourth version, says that 'Aging is a disease … Augmentation and enhancement to the human body and brain are essential for survival.' The goal is longevity, immortality even, where genetics, wearables and human–computer interaction create 'a transformation of the human species that continues to evolve with technology'. One moment of hope for transhumanists – their Independence Day, if you like – is the *singularity*, the moment in the future when machine intelligence becomes self-improving and begins to outstrip our human capabilities. A new age will dawn when we enter into a symbiosis with technology, and the human era as we know it will have ended.

And this is where the post-human cyborg becomes so exciting for people like Kevin. It is the potential to merge humans with super-intelligent machines. In this symbiosis the hope is that we'd be able to bring our complex human values and human control to the system and prevent AI being a danger. We will be able to solve many of the challenges that we face (poverty, food shortage, the energy problem, climate change – and, of course, human ageing, disease and mortality). Humanity will be emancipated from the flesh in which we suffer, with no more disability – we will be able to move beyond the social, religious and political bias and inequality that plague us, freed by technology.

'So, would you upload yourself if you could?' I ask Kevin. (One possibility the singularity might bring is uploading our minds to some kind of artificial super-intelligence substrate, in a sense swapping the fleshy wetware of our current bodies for something less prone to glitches and crashing – it would be the culmination of the cyborg project. Even as a hypothetical, I find the idea monstrous.)

He smiles, 'Yes, I would.' He then acknowledges that there are huge compatibility issues. As we don't yet fully understand how the human brain works, it will be a challenge, but Kevin has already proved that we can communicate with machines during Project Cyborg 2.0 – and then he offers up answers to some of the biological and technical challenges (there might even be advantages): transistors fire many times faster than our biological neurons, and while signals travel through the nervous system at around 100 metres per second, in an electronic system information can travel at the speed of light. And we wouldn't be limited by human anatomy, so a machine substrate could be much bigger than a brain.

I'd been trying to get Kevin to say something about the downsides or risks of the Project Cyborg experiments – part of me wanted him to acknowledge the loss of humanity I'd felt when assaulted by wires and tubes in hospital: the disconnection from my body, the risk of infection and danger of surgery, but he wouldn't bite.

'Speaking as a scientist, I didn't feel like there were negatives with the experiments we did. Firing the array into the nervous system felt to me like a perfect marriage. The fibrous tissue of the scarring holds the electrode in place – you end up with a better connection. I know it's strange. I can only really think of it scientifically.'

Kevin then describes how exciting it would be to have a nervous system that could reach outside the boundary of the body, into a network, or even link up separate human minds, with the chance of near-instant human communication and none of the hidden nuance or contradictions of spoken language – we could understand each other perfectly. (Whether understanding each other perfectly would be a good thing, I'm not sure, but I don't say anything.)

'In fact, joining my brain with another brain is an experiment I want to try,' he says.

Kevin's talk of uploading himself, or joining himself to another person, comes with the excitement of scientific exploration. In a sense he's made it his job to think this way. But the transhumanist wish for immortality through human transformation makes me feel a little appalled. I've been close to death, I might have been dead, and would never want to return to the pain and loneliness of that. Perhaps I should be desperate for the chance of immortality if it meant I didn't have to relive that sort of suffering. And yet I also have a feeling that death might in some way be important for a meaningful life – there's a comparison on the tip of my tongue: an immortal life is like a story without an ending, no part of the story will hold any meaning without the context of a conclusion. I can't quite articulate this to Kevin, sitting in his living room, and I suspect it won't stand up to any scientific scrutiny he might bring to bear.

I know I look at the cyborg and transhumanist dreams from a certain point of view – as someone who has had no choice but to become dependent on technology – and I can only apply my less-than-perfect relationship to it, and the anxieties that come from relying on a machine. Transhumanists talk of freeing themselves from the body, of transforming, and I raise an eyebrow. All I see is rubbing and sores, oozing body fluids and infection; the frustrations of not being as mobile as I would like, the anxiety of dependence on a machine. I can't imagine the pain, anxiety and frustrations of being a sentient being uploaded to a hard drive – what it would be like never again to feel rain on my face, or the visceral feeling of my family when we all bundle together on the sofa for a hug. In the future, being without the machine part of us, because it is broken or has crashed (or has given us an infection), might be wrenching in ways we can't yet imagine, especially if it is a super-intelligence and has let us access unimaginable realms of experience and freedom.

And then we're in the car and Kevin is waving us off. It's a long journey back to London in the rain and the traffic. Andy and I chat, but in the silences I turn over the conversation I'd had with Kevin. One part of the Project Cyborg experiments that he'd talked about – the part, he said, laughing, the press had found too weird to cover much – was when he connected through the BrainGate implant to the nervous system of his wife, Irena, achieving the first direct, purely electronic communication between two humans. His motor-nerve signals had travelled to her brain and caused her to feel 'lightning running from her palm up the inside of her second finger'. In one of Kevin's books, *I, Cyborg*, he'd said their marriage had looked to be on the rocks, but after the experiment they were closer than ever, 'having experienced something that no couple before us had experienced'. Given the opportunity to connect to another human, Kevin hadn't opted for one of his research assistants, but for his wife.

Even though many in the cyborg and transhumanist projects make huge imaginative leaps into the future that seem a little unfeeling and scientific, apocalyptic even, a lot of them would argue that a future human–machine symbiosis need not necessarily make us any less human; indeed, it might deepen our connection with each other and the natural world. And that's hopeful – many of their dreams are also the dreams of the disabled, so I'm glad they are out there pushing the boundaries of what is possible, ethically and practically.

But it's easy to think they are a little bonkers when they talk about these advances being just around the corner. I can't see it like that, when we struggle even to solve the problem of satisfactorily attaching a prosthetic leg. Perhaps this makes me too pessimistic, and technology will advance to provide unexpected answers. It has done so many times before; even if it takes a little longer than the futurists predict, we tend to get there in the end. And I suspect it's likely to be the disabled who will continue to be at the vanguard of testing out what is possible in this human–machine symbiosis – bodies already damaged enough to take the risk on.*

In the process, and if my own experience is anything to go by, there's hope that we won't lose what makes us human. For the most part, I don't feel any less human because of my prosthetics. I don't feel cyborg or robotic, either. What I feel is lucky to be able to walk, and stand, and pick up my children.

*

---

*Peter Scott-Morgan, a scientist who was diagnosed with motor neurone disease, has used surgery, robotics, augmentations and AI to become a cyborg and challenge the terminal diagnosis he was given. His deteriorating body becomes the laboratory on which to trial cyborg technologies that might prolong his life. The Channel 4 documentary *Peter: The Human Cyborg* tells his story.

During the weeks after meeting Kevin there remained a slightly unsettling feeling. It was this thing about the body being a suboptimal package that needs upgrading – I found the idea we should use technology to overcome death most troubling. I couldn't answer why I didn't agree with these ambitions. Then, late one sleepless night, some connections deep in my brain must have aligned and I remembered what I'd been groping for and picked up my phone, googled it and found the recording.

It was an episode of BBC Radio 4's *Desert Island Discs* that I'd heard repeated the previous summer. Dame Cicely Saunders, the founder of the hospice movement, is being interviewed and towards the end is asked, having looked after so many people at the end of their lives, how she would want her own death to be – would she want it to be speedy and painless? 'No,' she replies, 'I would like to have time to say thank you, one needs time to say I'm sorry, one needs time to sort out something of yourself, of what really matters, until perhaps you can finally reach the place where … you can say, *Well, I'm me, and it's all right.*'

Monsters

I am transforming: skin turning pallid white, teeth sharpening to fangs, hair hardening into spikes. My eyes darken and stretch upwards and my cheekbones protrude and my brows crease down into a grotesque scowl. I am a monster. I am quivering with fear. The others – the soldiers – are lined up, and we walk the corridor and I go to my spot on my own and wait until it is my turn. The moment nears with a beat I've learnt and dread. And then I am creeping through the dark, stepping my feet and clawing my arms in an exaggerated parody of the creature I'd been told to summon. But there is no feeling and it is lifeless. I am lit in a ring of intensity, and through the halo and brightness is the gaze of the audience and I am petrified. In the heat of the spotlight the paint on my face is a hot skin. I look through this mask, feeling nothing of what it is meant to mean; I can only be a ten-year-old boy. I attack *Soldier 3*, and *Soldiers 4* and *6* and *1*, batting away their cardboard shields, and they fall beneath me. I have to return later and again, and then *Beowulf* slays me.

After the make-up had been wiped from my face and I had changed out of the costume, I met my parents in the hall and wanted them to stop talking to the teachers and take me home, and I felt ashamed and confused. I had wanted to be *Soldier 5*, but was given the part of *Grendel*.

The original *Beowulf* texts used many of the Old English phrases for the monstrous to describe Grendel. One that's been up for grabs with modern critics and translators is *wiht unhælo*. While *wiht* is normally 'creature', *unhælo* is less certain. It's been interpreted as 'evil' or 'unholy', so Grendel becomes the 'unholy creature'; but *unhælo* turns out to be the closest word Anglo-Saxon had for 'disability', describing the disfigurements of disease and congenital conditions. With this reading, Grendel becomes 'being of sickness' or 'unhealthy being'. It's another myth in which perhaps the inspiration for the monster wasn't some magical beast, but a disabled person, cast out and made evil – a figure of fear and loathing.

The monstrous has been many things in the past, but mostly it is what we don't understand and what is frightening – distant lands and the foreigners who lived there, at the unexplored edges of the map where we wrote *here be dragons*; it's those who behaved or looked different, the lesbians we turned into witches or the deformed babies who became omens; it's the technology we couldn't yet comprehend – it was a metaphor for our anxieties. We still create monsters. They're the heinous criminals and natural disasters in the news headlines. Monster storm. Monstrous virus. The monstrous serial killer or paedophile. We turn threats to our culture and society into monsters, which can then act out our fears at a safe distance. It's a way of separating from ourselves what is threatening, making it *other*, but it's also a way of casting out the monster that we see when we hold up the mirror to ourselves.

To become *Grendel* was too difficult for the ten-year-old me. During rehearsals, in those brief moments when I did get out of myself, did act, it felt dangerous, as if I might lose a part of myself. In the end I couldn't bring anything to the performance, and the director – an English teacher who took me to one side for a pep-talk after the dress rehearsal and desperately tried to save his production – shrewdly decided to use a strobe light over most of my scenes, so the awkward ten-year-old with tears in his eyes was frozen in flashing stills of lamely wheeling arms and falling soldiers. I never went on a stage again and, from then on, I'd be found in the school-play programme as *Lighting assistant* or *Backstage 2*.

It felt the same after I lost my legs. I was the ten-year-old boy again, turned monstrous and on a stage I didn't want to be on – the lead in a play I didn't want to be in. Everyone was watching to see what would happen, to see the transformation I had made. The visitors, the letters and cards, the press articles, the knowledge that everyone was asking *How is Harry doing?* put me at the centre of a very hot spotlight. I was *Grendel* again, a 'being of sickness', with a body disabled. I wanted everyone to leave me alone – for the play to end.

I've searched out *Grendel* and *Quasimodo* and *Franken-stein's monster*: 'a figure hideously deformed and loathsome; I was not even of the same nature as man'. Franz Kafka's Gregor Samsa – a travelling salesman waking, transformed, from anxious dreams in *The Metamorphosis* – might be the closest to what I experienced. To wake one day as a monstrous insect, stuck on your back, spindly legs thrashing 'helplessly before his eyes', unable to roll over; and the strangeness of not being who you used to be, turned into something that sets you apart from those around you.

There were the practical implications of my disability that set me apart. The one that seemed easiest to reconcile was that I couldn't do the job I loved any more. Although there were short-lived fantasises that I might be able to go back and be useful, deep down I knew I could never be a soldier again. How this felt, more than anything, was embarrass-ing. I was meant to be lucky – luck being a basic attribute of any good soldier – but I'd made a mistake, unlucky or not, and that was profoundly embarrassing, in a professional sense. In time I got over this – it was even a release; I could do whatever I wanted now. But embarrassment went deeper, linked to other more difficult aspects of who I am, and this took longer to come to terms with. I was embarrassed by the changes in my body (yes, by what it meant I couldn't do, and by the dependence I felt on others), but mostly by how it looked: by being in a wheelchair, by the unnatural way I walked on prosthetics or, as a friend once said to me, 'You walk like you've got a carrot stuck up your arse.' All good, fun teasing at the time and I laughed, but it was a stake to the heart I'd never be able to forget. I'd worked so hard to make the way I walked look normal. The image and character I had projected to the world were gone – and it was all tied up with the anxieties many twenty-six-year-olds would have: will anyone fancy me?

If embarrassment is the feeling we get when we violate the particular persona we want others to see, I was awash with it for many months, years even. It did fade as I became comfortable with my new body and the experience of the world it gave me, and as I discovered that I was accepted – and could be loved. But embarrassment was tinged with shame, and that has lasted longer. While embarrassment is all about those personal standards we set for ourselves, shame is what we experience when we feel we've failed to live up to a shared standard of the society or group we are a part of. If there's something about us that doesn't fit with what is considered normal and natural, we fear we might be thrown outside the group. To where the monsters are banished.

Being ashamed of myself was all tangled up with not being able to show any vulnerability. I was a soldier, and being vulnerable (especially physically) was weakness. I'd been in an environment where it was normal and acceptable to say 'man up', or 'grow a pair', or 'go sign a backbone out of the stores' to someone who wasn't performing as they should, even to yourself under your breath. There were lots of these motivational sayings – 'pain is just weakness leaving the body' – some of them unrepeatable, and they also had a lot to do with shaming. At the time they seemed appropriate, part of the military code or game we'd signed up to. There was a line at which it stopped, when someone genuinely needed compassion, but the bar was high. It can be hard to defend this mentality to someone who hasn't been in a similar environment, but when you are in a team that has to go into combat, which might be the absolute aggression of a hand-to-hand gutter fight to the death, you have to be more ruthless than the opponent. There is little room for weakness when the result of not winning is second place, and second place is death.

This might actually have been an advantage during the physical and mental challenge of early rehabilitation. Just putting on the prosthetics every day, despite the pain, with the discipline and determination of someone who refuses to show any vulnerability or weakness, perhaps leads to a faster recovery. It may still stand me in good stead, helping me to continue the daily routine of donning my legs, despite the discomfort and the little voice saying, *Have a day off today*; I have a mantra to say against it: *Come on, get a grip of your body, Harry*. But in so many other ways it was damaging.

I don't want to overstate this, as it wasn't my whole experience, and many things can be true at once. My identity was in flux. We so often think (or society tells us) that you can only be one thing, when actually you can be simultaneously getting better, working out a new life, falling in love, having children, shedding the physical trauma of injury and finding a new appreciation of life you'd never felt before, and still have a small part of yourself that is conflicted and ashamed.

At its heart, shame is not feeling good enough – of not being worthy of connection. I'd felt shame as a ten-year-old

Grendel. I was terrified I'd lose the connection to the people I loved, but I didn't really understand that at the time. And I'm not sure I understood it much better when I suddenly acquired a disability fifteen years later. The shame of being a monster made me want to shrink away and disappear. With time, I came to accept myself, but the shame remained, shrivelled in a corner.

I'd heard of disability hate-crime but had never experienced it. It felt distant and someone else's problem. In truth, and it's hard to admit, when I heard a story of discrimination or read an article about how this sort of abuse was on the rise – up 12 per cent in a year – I might even utter one of those sayings to myself: *grow a pair.* But that internal voice had more to do with my own self-preservation than with any individual out there, hidden in the statistics. It probably went hand-in-hand with the fact that I didn't really see myself as disabled, and still found it hard to talk to someone else who was disabled (just as 67 per cent of people in the UK do). Part of me was who I used to be and who needed to be invulnerable, and part was still working out where I fitted in.

Then one evening I was walking into Clapham Junction train station, via one of the side entrances, and a middle-aged woman was walking towards me. She looked me up and down and, while sometimes people might smile or even say something nice, she said, 'Serves you right.'

'I'm sorry,' I said, thinking I'd misheard her.

'Serves you right,' she said a little louder and then was gone with her shopping bags.

I tried to excuse it, to explain it away. I put it down to her having a bad day or being unwell – or maybe it had been a political comment on the conflicts I represented. But the shame flooded back and I couldn't shake the unsettling feeling it gave me. As much as my partner told me not to worry, the words instantly made me utterly conscious of my disability and I felt shame deeply pervading me, in a way that is hard to describe. It made me shake and feel hot. I kept thinking I had misheard; but I was a monster again. It made me think again about those statistics – what it must be to feel this every time you leave the house.

That moment gave new perspective on my disability. Somehow *injured soldier* had set me apart – a special case where the issues of the disabled didn't count. But as time passed and I left the veteran community and all the military ethos that went with it, I found I had something in common with people I never would have had, if I hadn't been injured. The stories I heard of disabled people who had suffered hate: had eggs thrown at them because they had cerebral palsy; or were verbally abused because they had Down's syndrome; or were terrorised in their homes by gangs of teenagers for being dependent on an electric wheelchair – the lasting impact this has on confidence and mental health now made sense. I also knew that the gulf between my experience of disability, which assistive technologies can go some way towards fixing, and those with other severe mental and physical disability was so large as to make comparisons meaningless – my experience was merely the tip of the iceberg. Yet 'Serves you right' had been a hammer blow.

What I began to realise was it didn't have to be a hateful or even negative comment that made me feel a little of the shame of the monstrous; just trying to enter a building or get on public transport that wasn't accessible could bring it on – a practical nuisance, but also a moment that marked you out as different. And, most strange, it could be a comment intended as nice or positive. It would go in waves. I'd go months without anyone seeming to give me a second look, then people would stop me in the street, or every time I went to a playground with my kids I'd have children asking me about my legs, or innocently pointing and laughing. I'd be nice and explain as best I could, trying not to alarm their hovering parents when the children kept asking, 'But how did you lose your legs?' and I explained what a bomb was. They were only three or four years old, and they didn't mean me any harm, but it strained my patience and (not always – it seemed to depend on the situation and my own mood as much as anything – but often) made me aware of my difference.

Or the time I was sitting with my partner at the edge of the playground, now at the blissful stage where you can watch your children on the climbing frame rather than having to hover below them as health-and-safety officers, and a woman brought her two boys up and stood in front of me.

'I'm trying to educate them,' she said. (Something like this has happened on a few occasions.)

The boys looked blankly at her and me.

I said, 'Hi, guys, nice to meet you.'

She said, 'Your legs?'

I replied, 'Yes?'

I didn't want to be unkind, or to embarrass her in front of her children, but there was something about the way she had encroached and made my prosthetics the most important part of me when I was with my family, talking to my partner on a Sunday morning. I felt like a tourist attraction to be gawped at. I didn't seem to be able to summon the energy to engage with her or give her sons the guided tour. But it was also more pervasive – I felt some of the shame I'd felt after I'd met the woman in the train station.

I hope I was compassionate and tolerant before I was injured, but I had no way of seeing my true privilege. I had everything given to me on a plate. And yes, I'm still privileged, and I've carried that privilege into my life as a disabled person. I will never know what it feels like to be racially abused, or persecuted because you are from a minority group. I will never know what it feels like to be a black person and have someone lock their car doors as you walk past; or how it must feel to be told to go back to your own country when this is your home; or assaulted for kissing your same-sex partner on a bus. And I know there are probably inbuilt prejudices I carry and can't see, and may never be able to shed. But if there is one gift that my disability has given me that stands above all else, it is just the smallest window on what it is to be *other*.

I meet Andrew outside a London underground station, and the first thing we notice about each other is we have the same prosthetic foot – a blade that curves out behind the leg and sweeps forward to the rubber square of grip that touches the ground. By rejecting the space normal human anatomy fills, the bend of carbon fibre can generate more spring and be more efficient. It's an unusual leg to use every day, as it's designed for running. I've never seen anyone else with it. Nor has Andrew. People don't use it because it's hard to balance on – what it gives you in mechanical energy is paid for in stability – and also perhaps because of its weird and non-human look. Andrew likes it because of that. I'm more conflicted; I hate the way it makes me stand out. But the way it feels to walk on is too good to go back to something more conventional – function trumps appearance for me. Having the same leg makes me feel connected to Andrew in a strange way I hadn't expected, and we are laughing and joking.

We're meeting Sophie here. She's making a new leg for Andrew and has come to London for a fitting. As we wait and chat, someone comes up and talks to us, wondering about an uncle who is having an amputation, and how much it all costs and where we get our legs. Then Sophie is bundling out of the underground station with bags of bubble-wrapped objects and a backpack, and we are walking to where Andrew trains at the London Dance Academy.

Andrew had a motorcycle accident nineteen years ago. 'My bike instructor said never get on your bike if you are drunk or angry,' he says as we walk. 'I'd had an argument with a friend. I fell and my leg was crushed under the bike.'

They saved his leg and he managed the pain with painkillers and antidepressants. The doctors had talked of amputation after the accident, but it was only when he took up poledancing fifteen years later that he thought of it again. He'd been looking for an exercise he could do with his damaged leg. As soon as he tried pole-dancing he was hooked. It was new freedom for him. He could move without the leg getting in the way, but it also made him more aware of his disability, the cane he used, and the new physicality triggered flare-ups of pain. So he had his leg amputated below the knee. The rehab was hard, Andrew tells me, but now he can do far more and is in less pain.

At the Dance Academy we shuffle past people waiting for a class and into one of the small studios – mirrored walls and a pole at the centre. Andrew strips to his briefs and removes his prosthetic leg and peels down the liner. When we met I'd noticed his tattoos, and now I see they stretch blue-grey from neck to ankle – his stump is almost entirely covered with swirls and dots (he later told me the surgeons had done a great job of marrying the patterns when they sewed him back up) – and he is a striking presence, chalking his hands and hopping over to the pole. Sophie is in the corner, opening the bags and prepping the new legs; there are sockets and pieces of wooden carving laid out around her, a cogged rod that looks like a section of engine and a shaped rib of carbon fibre.

'Can I ask you about your tattoos?' I say. 'I'm not saying they're unusually—'

'You can't offend me, Harry,' Andrew says and laughs. 'I get that I don't look like everyone else. I'm very proud of that.'

While he stretches he explains. 'In my teens I started to think about tattoos and piercings, but it was a no-go area for my parents. I had terrible acne all over my body, on my face, chest and back; I had a bad relationship with my skin. When I moved to London in my early twenties it was starting to clear up. I was on this drug that wrecks your body but clears up your acne, and when I finished that I went and got this one little tattoo.' He has a hold of the pole now. 'One of those arm-bands that were trendy at the time. The tattoo artist turned out to be a pioneer of black tribal-work resurgence. I just booked another appointment and another appointment. I didn't even know what we were going to do; we just made it up on the day. As the tattoos appeared, I suddenly started to like my skin. So I had this bad relationship and I was turning it into something that I really liked.' He smiles. 'Now I say it, that's the same sort of thing I did with my damaged leg.'

'I love the tattoos,' Sophie says from where she is preparing. 'It's one of the reasons I wanted to work with Andrew.'

And then he is pulling himself up to the top of the pole. And he starts to shift his weight into a series of different moves that twist and ramp and swing off the pole.

I'd looked up a video of Andrew winning the drama category at the World Pole Sports Championships held a few months earlier. In the clip he comes onto the stage leaning on a crutch, throws it away, falls and then rises up onto the pole to tell the story of his amputation and rehabilitation. It is beautiful. It also looks effortless, with the distance of the camera and screen. But here in the studio it is totally different. I can see the strength and effort required, and can hear the skin of his hands squeaking slightly as he grips the pole and holds his body upside-down and his legs outspread, the pole vibrating under the forces he sends through it as he pendulums into another move.

He'd told me on the walk over that he'd hated his damaged leg. 'I hated the visual of it. It was twisted and deformed – maybe not to other people, but to me it was monstrous and wrapped up in the pain of everything it represented. I had to use a walking stick; I couldn't just go out and buy a pair of shoes, because of the heel. Sixteen years I lived like that.'

Now up on the pole, warming up with a few moves, his stump is very eye-catching. He can send it through gaps between other body parts that an able-bodied dancer couldn't, and the weighting and balance are captivating. Then Sophie is ready and calls Andrew over, and he sits in front of her and they begin trying on the new leg. Andrew is genuinely excited. It's the first time he sees it.

'Wow, I'm blown away,' he says. 'This is so cool.' (He says *wow* a lot over the next forty minutes.)

The leg is still being made – this is a fitting to check the comfort, shape and function are correct – and the surface of the limb is unfinished carbon fibre, dusty from grinding and shaping. There is something very *other* about the limb as Andrew pulls it on, and a sense of what it might be when it's complete. It's also very 'Andrew'. The material is the same tone as his tattoos and it sweeps from his calf, narrowing like an ancient fire-blackened whale bone, widening to a clubbed end. When Andrew stands up, it looks more like a distorted, very high-heeled stiletto, and a little like an animal hoof, fuse-jointed and at once from the past and the future.

After the amputation Andrew had dreamt of having a

prosthetic that didn't look like a leg and he watches it in the mirrors. 'The way it's like a hoof is so beautiful,' he says with real joy. 'Why doesn't everybody want hooves?' He takes a few careful steps. 'I can't believe how much it feels totally like part of me.'

He lifts himself up onto the pole and starts testing it out. Suddenly he is complete, and the strangeness of the amputation is gone. At first glance you might not notice the limb as it sweeps through another arc – less strange than seeing the amputation – but then Andrew is near the top of the pole, upside-down and holding a position with the leg extending out and frozen for a moment. Even unfinished, the limb that Sophie has made is exciting, following the curves of Andrew's tattoos, part tribal war club, part animal, part sleek high fashion.

Sophie runs the Alternative Limb Project. I'd been to meet her in the workshop on an industrial estate in Lewes. We talked about her work as she mixed pigment into silicone, feeding it between the rollers of an electric mill, which worked the colour through. She was making an ultra-realistic cosmetic limb for a farmer who'd lost his foot in a motorcycle accident. Beside her was a sheet with colour-match notes made from his skin and photographs, and a cast of his good foot. She was making an exact mirror, and would add the nails and hair, layer those blue-green translucent veins beneath the surface until no one would take a second glance when he went to the beach. This is Sophie's bread and butter. To demonstrate how lifelike the prosthetic would be, she kicked off her plastic clogs and pulled on a pair of silicone slippers over her socks. They are exact copies of her feet, so real – with all the uniqueness of each individual toe – that it was like looking at her feet, except that you could still see her socks disappearing inside them, and they were too big. She called them her hobbit feet.

I told Sophie I felt something of the uncanny valley in these ultra-realistic prosthetics.* They are just too dead-looking for me, even the incredibly lifelike ones she can conjure. Sophie didn't agree; these types of prosthetics could be important for people who felt the loss of their limbs keenly and wanted as close a visual replacement as possible – for some people, not standing out, feeling and looking complete, is more important than function.

It was odd that half her work focused on creating absolute copies of the human form, and the other half was playing around with what alternatives might be possible. And the

---

*The uncanny valley is the idea that as human-like creations become more human, we feel increasingly positive towards them, until they are too human-like. It's illustrated in a graph: on the x axis is machine–human likeness, on the y axis is likeability. The line of the graph goes up as a creation gets more human and we like it more, but there's a point when the resemblance is suddenly nearly, but not quite, human and we feel revulsion. The uncanny valley is a sharp dip in the graph. It's applied in both prosthetics and robotics – make them too lifelike and they are unnerving, reminding us of disease and the un-dead. As technology lets us create ever more believable humanoid robots and incredibly realistic visuals in film, gaming and VR, the concept has jumped from academic to popular culture.

workshop was filled with found objects that inspired her, wooden carvings, metalwork and a cabinet of curiosities with old prosthetics belonging to Chris, the prosthetist who worked with her. Sophie searched scrap heaps and junk shops for inspiration – she wasn't sure what to do with a wooden canoe propped up outside the entrance. It was more artist's studio than prosthetist's workshop, and camouflaged among the hanging plant baskets was one of her alternative limbs, titled Vine, an upper-limb prosthetic that, instead of a hand, was a long, articulated tentacle-like appendage, which curled around objects to pick them up. If the cosmetic limbs paid the bills, then the alternative limbs were what really interested Sophie.

She showed me some of the limbs she'd made, flicking through pictures from photoshoots. First, a number of lower-limb prosthetics made for the pop-artist singer Viktoria Modesta: a black Spike – no foot – just a single shiny shard with a sharp point to stand on; one glittering with crystals and rhinestones for the 2012 Paralympics closing ceremony; and one for a Rolls-Royce campaign with a mini sparking Tesla coil inside; then Snake Arm, made for a hand amputee, an ultra-realistic cosmetic arm with a snake wrapped around it and crawling, rather unsettlingly, through a cavity in the wrist; another, playing on the tension of real and unreal, was Materialise, whose lower half of the arm was skin-like with two perfect fingers, while the upper half was made from interchangeable sections of rock, earth, cork, oil and moss – each related to a different emotional and spiritual part of the amputee's personality; then she flicked up Cuckoo, a beautiful leg made for a contemporary dancer, carved from cherrywood, and a little steampunk, which was also a time-piece with a watch and a cuckoo that popped from the knee.

There were dozens of different limbs. All the pictures showed the amputee in a staged photoshoot, missing body part now changed into something new and unexpected. Sophie said the limbs might be about the character of the person, or a journey they'd been on, or an aesthetic she wanted to explore. Some were highly stylised and high-fashion, others more playful and personal. Sophie was now in conversation with a woman in the US who wanted to carry her mother's ashes in her prosthetic hand – her mother had spent a lifetime blaming herself for her daughter's con-genital limb difference, and this was part of the forgiveness and acceptance. I found this heartbreaking. And she had a project going with Darlington Railway Museum to create a prosthetic leg with a model railway running around it: a miniature train on a track that passed through tunnels in the limb. She was toying with the idea of using a vaping cigarette to create the puffs of steam.

After studying special effects at university and working for a few years in film, Sophie had found her way into cos-metic prosthetics for amputees. One client, a little girl who came to see her every year, wanted cartoon characters or pictures of her family on her prosthetic. The girl had been in a pushchair when a bus mounted the pavement, killing her grandmother, scarring her mother and damaging her leg beyond repair. Then one year she came in with a drawing of a leg with little drawers to keep her things in. For Sophie, this wish to personalise a limb was an interesting aspect of rehabilitation that hadn't been explored properly and led her to start the Alternative Limb Project. In the beginning she'd enjoyed making beautiful, sexy prosthetics, like the ones for the pop star, but she felt there was a range of other emotions to explore: ideas of identity, body image, modification, transhumanism.

As I'd left her workshop, Sophie had handed me a book she said she'd found useful. I'd thrown it on the passenger seat of the car and flicked through it when I stopped in a service station on the way home for coffee: *Staring: How We Look* by Rosemarie Garland-Thomson. It was an academic text about what it is to stare. While most theories of staring have focused on the power, ownership and violence of the starer, this book suggested it was not so simple, that there was an important change to be made in the way we think about staring.

Reading through the introduction reminded me how much I'd hated being stared at after injury; I would wait until no one was around before getting out of the car, anything to avoid people seeing the ungainly prosthetics – all part of the embarrassment and shame. I still find it uncomfortable, the sense I get that someone is turning to take another look at my legs when I pass them in the street. I used to spin round to catch them at it, challenging them with my own stare: *How do you like that?* I've become better at being stared at now, open to engaging with the starer with a smile or a hello.

We stare at a car crash, or some natural optical illusion, or ourselves in the mirror, but when we stare at another person it creates a connection between starer and staree. The stare is bad when we look at a car crash, or a photograph of a migrant washed up on the beach, or the disfigured human, and think: *Well, at least that's not me.* But the book suggested ways in which this connection could be positive: when the stare is one of empathy; when it's transformed into a compassionate transaction between starer and staree. We look for too long, but think, *That could be me*, and it makes us take action. And the book made a case for broadening what is expected to be seen in public – an acceptance of the diverse human community – so that those who are traditionally stared at, those with monstrous bodies, do not *hide or allow themselves to be hidden.*

It's very hard not to stare at Andrew on the pole. But then that's what he wants. It's taken him a few minutes to get used to the new leg. There's been a slight strain on his face as the prosthetic clangs uncomfortably into the pole. I imagine he's having to adjust finely balanced moves to the new weight and length of the leg. He quickly improves, though, and strings together a sequence. 'Oh my God, I feel like a superhero,' he says after completing a section of routine and standing, chest heaving, beside the pole, looking at his reflection in the mirrored walls of the studio.

He wants Sophie to make him a limb that extends his abilities, and the toe end of the sleek 'hoof' has a notch that clips onto the pole. Over the next half hour Andrew returns the leg to Sophie with a few comments, and she disappears out into a courtyard and grinds away at the notch until it can grip the pole. It's very like any conversation you'd hear in a limb-fitting centre – *take a bit off here; it's rubbing here; what about widening there* – except you'd never see this limb in a medical setting.

When Andrew has a breather, and Sophie again takes the leg into the courtyard to be adjusted, I ask him about how important disability is to his performance.

'I turned my disability into an identity,' he says. 'I have this character, my pole-performing persona, *tattoo_pole_boy* on Instagram. Everyone knows I'm an amputee and I've used that to my advantage. My disability creates a strong image. It's hard to stand out in a sport like this. It's led to opportunities – interviews and articles. That's the thing with Sophie's leg, I want to push it to that next level.' Sophie has brought the limb back and Andrew is pulling it on again. 'It's also about my look. Anything that's different from a normal leg excites me. And the odder it looks, the more human I feel. My biggest fear in life is blending in. It frightens me. I want to be noticed ... I don't want to shove it in people's faces ... but I want to stand out.'

After a final test, Sophie calls him over to try the other limbs, and Andrew takes off the pole leg. 'I don't want to give it back,' he says.

While the first leg is very much for Andrew, made with his creative input and for his performance, this second limb is Sophie's concept. Different attachments can be clipped onto a socket. She has two here today: one brass leg, taken from a piece of antique furniture, joined to a bespoke wooden carving that she's worked on with a carpenter – it's an intricate lattice that fits around the brass and looks like the prow of a miniature man-o'-war. The other limb uses the gearbox shaft I'd seen her unpacking earlier, and the cogs and machined metal make for a very different object. Andrew tries them both, and Sophie jots down measurements and they talk about next steps. She'll take them back to the studio and continue to work on them before the next fitting.

This is often how Sophie works. She finds interesting people who are up for exploring their bodies in new ways, applies for funding from arts or health organisations and develops a project. Andrew gets the alternative limb of his dreams, in return for modelling some of Sophie's imaginings – and while it is a collaboration, with Andrew bringing his individuality and opinions, the concepts that Sophie explores are more likely to end up as photographs in a magazine, or as displays in museums and galleries, than being used day to day. Reaching an audience is a crucial part of the project. She's had alternative limbs in exhibitions across the world. More than anything, Sophie wants to promote positive debate around disability and to celebrate body diversity.

Watching Sophie and Andrew talking through the technical problems and aesthetic of the legs reminded me of the 1985 essay I'd read as an undergrad, 'A Cyborg Manifesto' by Donna Haraway (a professor of the history of consciousness at the University of California), who dreams of a monstrous world without gender. In the essay – a critique of the traditional feminist discourse, that is now a set text on many university programmes and has almost cult status – Haraway says the Western cultural traditions of patriarchy, colonialism and naturalism have created oppositions that dominate women, people of colour and everyone who is considered to be *other*. She argues that technology can be used to address the domination of these traditions and provide 'a way out of the maze of dualisms in which we have explained our bodies'. Haraway's cyborg is the entanglement of humans and technology (now so intimate it's impossible to unravel), which breaks down the rigid boundaries of animal, human and machine.

The ideas in the essay have been used to give new meaning to many academic fields, often overlooking some of the original concepts of Haraway's cyborg, but harnessing her powerful imagery. And as Andrew stands there now, with a gearbox shaft as a leg and the clunk it makes on the studio's wooden floor as he steps, it feels easier to imagine a future where we might construct our bodies with tech, choosing to express our identity, sexuality and gender more freely; where a human–machine hybrid might make the discrimination and domination of traditional Western categories obsolete; where the boundaries become confused and less important (and that's not necessarily making more categories, but to rid us of categories altogether). The possibilities are inherent in all the technologies we have merged with, the information networks we use to form new communities, and the augmentations and replacements to our bodies that reduce physical difference. And with Andrew having a last try on the pole with his new limb, reflected infinitely in the mirrored walls of the studio, the *monstrous world* does seem closer.

The monster is being reclaimed. You're now just as likely to see a kind, compassionate monster in pop culture as a repulsive one: Darth Vader's amputations become Luke Sky-walker's, Captain Hook's becomes Hiccup's, the hero of *How to Train Your Dragon*. There are still all sorts of ways in which the monstrous gets sexualised (it has been for centuries) – the ultimate being the 'dating the bad boy/girl' narrative of monster movies: *King Kong* (1933), *Beauty and the Beast* (1991) and *The Shape of Water* (2017). But even in these examples you can chart the shift: King Kong is slain; the Beast gets to live but, disappointingly for most people watching, turns into a rather less attractive human to better suit Belle; only in *The Shape of Water* does the 'The Asset' stay monster at the end of the film (Elisa even becomes monstrous to join him; it leans heavily on *Shrek*). They all asked us to feel empathy for them, but only recently can we let the monsters be themselves.

You can see the monster taking hold across pop culture. Lady Gaga – who has used wheelchairs and prosthetics in her performance, and monstrous make-up and costume, and whose 2009 album was called *The Fame Monster* – has fans who call themselves *Little Monsters*. It's a community where they can celebrate their difference from mainstream culture, and the monster becomes a positive symbol of togetherness. This is a monster of creativity, resilience and compassion. By celebrating their difference as a community, they have formed a space in which any identity is acceptable. In the past, monsters were created by others; now, more than ever, they create themselves. Those who don't feel they fit into the constructed categories of 'normal' Western society – those who felt the shame of standing outside those categories, or the shame of suppressing a part of themselves to live within them – now have new ways of expressing and disrupting what it might mean to be human.

Technology can remake the body so that it appears less monstrous, reconstructing deformities, filling the absence of amputation, replacing the loss of a sense and reducing the physical differences between people. It can hide a disability so there is no stare. But it can also help us celebrate and express the monster in us, letting us move about the world and communicate with each other to form strong networks and alliances. And it can transform the stare into something positive – *Wow, your legs are cool* – and even let us transform ourselves into something new that goes beyond the human.

Many people feel we need the safety of clear-cut definitions, that we must choose to fit within them. And perhaps a society of the monstrous does have risks, perhaps it is a jungle of anarchy and a threat to much that is precious – there will always be dangers in the networks and bodily transformations that technology might let us create. But while so many of us don't happily fit within society's traditional categories, and with the inequalities they create, surely any way in which they can be broken down, and people can express who they are, is hopeful. In the age of the hybrid human the monstrous can be a celebration of difference. We all have a little monstrous in us, don't we?

The fitting is over, and I ask Andrew how he feels, wearing the alternative limb.

'It makes me feel incredibly powerful,' he says. 'Special and unique. I want to run out of the studio and down the street to show everybody.'

Golden Repair

I was sitting across the kitchen table from my partner. We were discussing cancelling plans to see friends and family, and a longed-for holiday we'd booked. The global pandemic had taken everyone by surprise and the prime minister had just told the country to stay at home.

On the table between us was the wedding present my brother-in-law had given us. Sometimes we filled it with fruit, but it was empty now. I'd always liked it. Almost half the bowl was unmarked white porcelain, and the rest was criss-crossed with golden-lined cracks, converging to a close-knit epicentre of crazed triangles. He'd bought the bowl, smashed it with a hammer and painstakingly put it back together using *kintsukuroi*, then packed it in its box.

Kintsukuroi, or Golden Repair, is a Japanese technique that uses lacquer mixed with gold to repair pottery. I'd looked it up and searched out old examples in London museums. A vessel is not thrown away when it breaks. The break becomes part of the history of the object, and the beauty and the time spent on the repair add to the reason the object is valued. The flaws and imperfections are embraced, turned golden by the liquid lines. In an age of throw-away consumerism, the bowl stands out in our house – it's more beautiful and interesting to me than the original ever would have been.

Later that week I lay in bed awake. Anxiety consumed me in the darkness. The busy road at the end of our street, normally a constant whooshing, was silent now and my thoughts had turned apocalyptic. I had little fear of the virus. I could see in the science that I would be unlucky to suffer serious illness. But my morning appointment at the limb-fitting centre had been cancelled and I was catastrophising: what if the economy collapses, what if I lose my job, what if the NHS fails? My imagination was spiralling: what if the power stops, how will I charge my prosthetic leg? What happens if it breaks or I get an infection now? What if the companies that supply my med-tech go bust? How will I find food for my family? How many tins of tomatoes do we have downstairs?

Everyone's lives were suddenly made precarious, and disabled communities would feel this keenly. I noticed the very small ways in which the pandemic affected my disability: wearing a face mask obscured where my prosthetics were stepping and I found I tripped more (and a friend told me her mask kept dislodging her hearing aid); I noticed the people I met for the first time on video calls treated me differently, not seeing my disability. (We all discovered video-conference calling, but I thought of the disabled groups, too vulnerable to meet in person, who had connected this way for years – leading the way with tech, again.) But the impact on the disabled would be far greater: isolated from their social networks and healthcare support, they would make up six of every ten people who died from the disease in the UK.

A siren warbled past and I sighed and turned over. The pandemic had ended my search for those who pushed at the boundaries of what it means to fuse the body with technology. Meetings and symposiums were cancelled and I hunkered down with my family. After talking with so many hybrids, I felt more comfortable with my disability. Even *disabled* felt okay – still not quite right, but language is imperfect. And I'd learnt that people had fought for that word, and the rights that came with it. I still liked *hybrid human* as a way to describe my experience, but the technologies I'd discovered remained far from being anything other than assistive – I was disabled, and always would be. Accepting this was a relief: being thankful for what I had, rather than striving for some unrealistic expectation of what might be possible, had made me happier.

The pandemic knocked some of this optimism. It made me aware of how dependent I was. The technologies that repaired me were so connected to the companies, healthcare providers and government institutions of a functioning society, and in the middle of the night it felt as if those foundations were shaking. It reminded me that I would remain dependent on the society I knew I was very lucky to be a member of. When I heard people say we had to let the virus run its course, that we were protecting the old while the young suffered, it made me wonder who gets left behind when the boat starts to sink. A society that doesn't look after the vulnerable isn't looking after anyone – I'd learnt first-hand that we're all just a moment from becoming vulnerable.

When I acquired a disability, I hadn't been chucked on the rubbish tip. Like the *kintsukuroi* bowl, I had been remade. This process of repair, and the time I'd spent on it, added to the reason I valued my life more. I had embraced my flaws and imperfections, those golden lines of my repair that were interwoven with the medical technologies I relied on. This was why I'd found myself feeling a little hero-worship for the therapists and doctors, the engineers and scientists I'd met along the way – they were the ones who had repaired me. I felt hopeful that science would give us a collective way out of the pandemic. Vaccines and therapies would be developed. I knew, just like so many of the technologies I'd seen over the last year, they wouldn't live up to the hype – there would be no silver bullet, and we would be turning the dial towards freedom and out of the trauma, but perhaps we would find a little collective Golden Repair in the process.

The light of morning shifted my anxiety and after breakfast I sat on the floor, staring through the wall as my children played around me. My partner was working upstairs. We would swap at lunch. I looked over at my son. He was wrestling with a cardboard delivery box by the front door. He put it over his head, walked into the wall, wobbled and then I heard his muffled voice, 'I am robot.'

I pulled the box off him, cut holes for his face and arms with the bread knife, squashed him back in and he walked round and round the kitchen table, grinning from the opening, arms sticking from each side and repeating, 'I am robot.' And my daughter laughed and goaded him and threw cuddly toys at him.

Bibliography

Some of the articles, books and websites consulted are given below.

Articles

Baird, Woody, 'Spirit Thrives After 41 Years in Iron Lung: Polio: Dianne Odell says her life is full of love, faith and family; that you can make your life good or bad', *Los Angeles Times*, 13 October 1991.

Baumgaertner, Emily, 'As D.I.Y. Gene Editing Gains Popularity, "Someone Is Going to Get Hurt"', *The New York Times*, 15 May 2018.

Benabid, Alim Louis et al., 'An exoskeleton controlled by an epidural wireless brain–machine interface in a tetraplegic patient: a proof-of-concept demonstration', *Lancet Neurology*, 18 (12), December 2019, pp. 1112–22.

Bhattacharjee, Yudhijit, 'Scientists are unraveling the mysteries of pain', *National Geographic*, 17 December 2019.

Biddiss, E. and Chau, T., 'Upper-limb prosthetics: critical factors in device abandonment', *American Journal of Physical Medicine &*

Rehabilitation, 86 (12), December 2007, pp. 977–87; doi: 10.1097/
PHM.0b013e3181587f6c; PMID: 18090439.

Buquet-Marcon, Cecile, Philippe, Charlier and Anaick, Samzun,
'The oldest amputation on a Neolithic human skeleton in France',
Nature Precedings, 30 October 2007.

Callaghan, Greg, 'The astonishing journey of surgeon Munjed Al
Muderis', *Sydney Morning Herald*, 18 September 2014.

Callaway, Ewen, '"It will change everything": DeepMind's AI makes
gigantic leap in solving protein structures', *Nature*, 30 November
2020.

Clark, Andy and Chalmers, David, 'The Extended Mind', *Analysis*, 58
(1), January 1998, pp. 7–19; www.jstor.org/stable/3328150.

Click, Melissa A., Lee, Hyunji and Holladay, Holly Willson, 'Making
Monsters: Lady Gaga, Fan Identification, and Social Media',
Popular Music and Society, 36 (3), 14 June 2013, pp. 360–79; doi:
10.1080/03007766.2013.798546.

Cockburn, Patrick, 'Polio: The Deadly Summer of 1956', *Independent*,
1999.

Drinker, Philip and Shaw, Louis A., 'An Apparatus for the Prolonged
Administration of Artificial Respiration: A Design for Adults and
Children', Harvard School of Public Health, 1928.

Erle, S. and Hendry, H., 'Monsters: Interdisciplinary explorations in
monstrosity', *Palgrave Communications*, 6 (53), 25 March 2020; doi:
10.1057/s41599-020-0428-1.

Eshraghi, Adrien A. et al., 'The cochlear implant: Historical aspects
and future prospects', *Anatomical Record* (Hoboken), 295 (11),
November 2012, pp. 1967–80; doi: 10.1002/ar.22580.

Finch, Jacqueline, 'The ancient origins of prosthetic medicine', *Lancet*,
12 February 2011.

Gallagher, James, 'Paralysed man moves in mind-reading exoskeleton',
4 October 2019; www.bbc.co.uk/news/health-49907356.

Gawande, Atul, 'The Itch: Its mysterious power may be a clue to a new theory about brains and bodies', *New Yorker*, 23 June 2008.

Glenn, Linda MacDonald, 'Case Study: Ethical and Legal Issues in Human Machine Mergers (Or the Cyborgs Cometh)', *Annals of Health Law*, 21 (1), Special Edition 2012.

Green, Rylie, 'Elastic and conductive hydrogel electrodes', *Nature Biomedical Engineering*, 3 (9–10), 8 January 2019.

Komorowski, M., Celi, L. A., Badawi, O. et al., 'The Artificial Intelligence Clinician learns optimal treatment strategies for sepsis in intensive care', *Nature Medicine*, 24, 22 October 2018, pp. 1716–20; doi: 10.1038/s41591-018-0213-5.

Lambert, Bruce, 'Dr R. Adams Cowley, 74, Dies; Reshaped Emergency Medicine', *The New York Times*, 1 November 1991.

Lawrence, Natalie, 'What is a monster?', University of Cambridge Research, 7 September 2015; www.cam.ac.uk/research/discussion/what-is-a-monster.

Learmonth, Ian D. et al., 'The operation of the century: Total hip replacement', *Lancet*, 27 October 2007.

Li, Xiao et al., 'Archaeological and palaeopathological study on the third/second century BC grave from Turfan, China: Individual health history and regional implications', *Quaternary International*, 2013, pp. 290–91.

McMenemy, Louise et al., 'Direct Skeletal Fixation in bilateral above knee amputees following blast: 2 year follow up results from the initial cohort of UK service personnel', *Injury*, 51 (3), March 2020, pp. 735–43.

Madrigal, Alexis C., 'The Man Who First Said "Cyborg," 50 Years Later', *Atlantic*, 30 September 2010.

Martiniello, Natalina et al., 'Exploring the use of smartphones and tablets among people with visual impairments: Are mainstream devices replacing the use of traditional visual aids?', *Assistive Technology*, 7 November 2019, pp. 1–12.

Max, D. T., 'Beyond Human', *National Geographic*, April 2017.

Mudry, A. and Mills, M., 'The Early History of the Cochlear Implant: A Retrospective', *JAMA Otolaryngology – Head & Neck Surgery*, 139 (5), May 2013, pp. 446–53; doi: 10.1001/jamaoto.2013.293.

Robertson, Adi, 'I Hacked My Body for a Future that Never Came', *The Verge*, 21 July 2017.

Santos Garcia, João Batista and Barbosa Neto, José Osvaldo, 'Living without the opioid epidemic: How far have we come?', *Lancet Neurology*, January 2020.

Talbot, Margaret, 'The Rogue Experimenters: Community labs want to make everything from insulin to prostheses. Will traditional scientists accept their efforts?', *New Yorker*, 18 May 2020.

VanEpps, J. Scott and Younger, John G., 'Implantable Device-Related Infection', *Shock*, December 2016.

Vita-More, Natasha, 'Transhumanist Manifesto', First version 1983; second versions 1998; third version 2008, fourth version 2020; www.natashavita-more.com.

Zeng, Fan-Gang, 'Trends in Cochlear Implants', *Trends in Amplification*, 1 December 2004 (re Alessandro Volta comment).

Zettler, Patricia J. et al., 'Regulating genetic biohacking', *Science*, 365 (6448), 5 July 2019, pp. 34–6; doi: 10.1126/science.aax3248.

Zhang, Sarah, 'A Biohacker Regrets Publicly Injecting Himself with CRISPR', *Atlantic*, 20 February 2018.

Books

Brown, Brené, *Rising Strong*, Vermilion, 2015.

Calhoun, Lawrence G. and Tedeschi, Richard G, *Handbook of Posttraumatic Growth: Research and Practice*, Psychology Press, 2014.

Garland-Thomson, Rosemarie, *Staring: How We Look*, Oxford University Press, 2009.

Godden, Richard H. et al., *Monstrosity, Disability, and the Posthuman in the Medieval and Early Modern World*, Palgrave Macmillan, 2019.

Haddow, Gill, *Embodiment and Everyday Cyborgs: Technologies That Alter Subjectivity*, Manchester University Press, 2021.

Haraway, Donna J., 'A Cyborg Manifesto: Science, Technology, and Socialist-Feminism in the Late Twentieth Century', in *Simians, Cyborgs, and Women: The Reinvention of Nature*, Routledge, 1991, pp. 149–81.

Hull, John M., *Notes on Blindness: A Journey Through the Dark*, Wellcome Collection, 2017.

Lyman, Monty, *The Remarkable Life of the Skin: An Intimate Journey Across Our Surface*, Black Swan, 2020.

Mayhew, Emily, *A Heavy Reckoning: War, Medicine and Survival in Afghanistan and Beyond*, Wellcome Collection, 2017.

——, *Wounded: The Long Journey Home from the Great War*, Vintage, 2014.

Moravec, Hans, *Mind Children*, Harvard University Press, 1988.

Näder, Hans Georg, *Futuring Human Mobility*, Steidl, 2019.

O'Connell, Mark, *To Be a Machine*, Granta, 2017.

Shelley, Mary Wollstonecraft, *Frankenstein; or, The Modern Prometheus: The 1818 Text*, Oxford University Press, 1998.

Solomon, Andrew, *Far from the Tree: Parents, Children and the Search for Identity*, Vintage, 2014.

Tracy, Larissa, *Flaying in the Pre-Modern World: Practice and Representation*, D. S. Brewer, 2017.

Wall, Patrick, *Pain: The Science of Suffering*, Weidenfeld & Nicolson, 1999.

Warwick, Kevin, *I, Cyborg*, Century, 2002.

Winters, Robert W., *Accidental Medical Discoveries: How Tenacity and Pure Dumb Luck Changed the World*, Skyhorse Publishing, 2016.

Websites

deepmind.com/blog/article/alphago-zero-starting-scratch

natashavita-more.com/transhumanist-manifesto/

thealternativelimbproject.com/

www.cyborgfoundation.com/

www.medtecheurope.org/wp-content/uploads/2020/05/The-European-Medical-Technology-Industry-in-figures-2020.pdf

www.ons.gov.uk/peoplepopulationandcommunity/healthandsocialcare/disability/articles/nearlyoneinfivepeoplehadsomeformofdisabilityinenglandandwales/2015-07-13

www.tedxbratislava.sk/en/video/jowan-osterlund-biohacking-next-step-human-evolution-dead-end/

www.youtube.com/watch?v=r-vbh3t7WVI&feature=youtu.be&t=5405 (Neuralink Launch Event)

A Note on Ableism

I've touched on tackling my own internalised ableism, and the ableism that permeates our society, in this book. We value 'perfect' bodies and abilities. Our world is designed for these people, and anyone diverging from the norm can find it difficult to navigate. If *disablism* is discrimination or prejudice against disabled people, then *ableism* is the system of values that places perfect bodies, abilities and productivity at the top of the pile (see www.scope.org.uk/about-us/disablism). It's an important distinction because it puts emphasis on the unconscious bias we all carry. I went from bedridden to being a wheelchair user to walking on prosthetics; I travelled along a spectrum of disability – from utter dependence to tech-assisted independence. And although my experience makes me more aware of the barriers and discrimination disabled people find in a world made for the able-bodied, my biases still exist.

My journey from dependence to independence – in a sense, being 'fixed' so I was more 'normal' – was thanks to the resources I had access to. So often, society funds and supports access to technology for some disabilities over others – my own experiences as a young white male soldier will differ radically from those of an older amputee from an ethnic minority, who needs an electric wheelchair costing tens of thousands of pounds but isn't eligible for a generous government funding pot. In this way, many of us are doubly 'disabled' by society. This intersectionality – the way different social, financial and political factors combine to create cumulative barriers – is important to remember when we think about ableism.

While technology *can* help us, we still need the world to do more to adapt to the needs of disabled people, rather than looking to them to 'adapt' themselves. We will always carry biases, but being aware of them is the first move to change. To learn more about ableism and the social model of disability, and how these issues play out, you might try starting with the following resources: head to YouTube to watch 'The Social Model of Disability' by the National Disability Arts Collection and Archive and Shape Arts; or 'Ableism and Language' by Social Work Technical Writer; read the book, blogs or listen to the podcasts at The Disability Visibility Project (www.disabilityvisibilityproject.com); read 'A Brief History of Disabled People's Self-Organisation', produced by Greater Manchester Coalition of Disabled People; or 'A Disability History Timeline: The struggle for equal rights through the ages', produced by NHS North West.

Acknowledgements

Thank you to all those who helped with inspiration, support and research: Dominic Aldington, Gus Alexiou, Andy Augousti, David Belton, Rik Berkelmans, Simon Bignall, Richard Bignall-Donnelly, Clothilde Cantegreil, Dani Clode, Michael Crossland, Hayden Dahmm, Alex Dainty, Sophie de Oliveria Barata, Christian Dinesen, Walter Donohue, Daniel Dyball, Stewart Emmens, Demetrius Evriviades, Aldo Faisal, Hugo Godwin, Andrew Gregory, Candace Hassall, Christopher Hastings, David Henson, Jamie Jackson, Julian Jackson, Douglas Justins, Rosie Kay, Angela Kedgley, Jon Kendrew, Pete Le Feuvre, Ruth Lester, Ross MacFarlane, Tamar Makin, Thomas Manganall, Emily Mayhew, David McLoughlin, Louise McMenemy, Akinnola Oluwalogbon, Jack Otter, Caro Parker, Jolyon Parker, Sophie Parker, Louise Read, Kate Sherman, David Smith, Emily Sparkes, Jennifer Sweet, Nigel Tai, Mark Thorburn, Jim Usherwood and Kevin Warwick.

Thank you to my agent, Matthew Hamilton, editor Francesca Barrie, and everyone at Wellcome Collection and Profile Books: Peter Dyer, Claire Beaumont, Teresa Cisneros, Louisa Dunnigan, Alex Elam, Lisa Finch, Lottie Fyfe, Mandy Greenfield, Jonathan Harley, Ellen Johl, Anya Johnson, Ed Lake, Nathaniel McKenzie, Niamh Murray, Angana Narula, Rosie Parnham, Patrick Taylor and Valentina Zanca.